Forensic Science
in
Wildlife Investigations

INTERNATIONAL FORENSIC SCIENCE AND INVESTIGATION SERIES

Series Editor: Max Houck

INTERNATIONAL FORENSIC SCIENCE
AND INVESTIGATION SERIES

Forensic Science in Wildlife Investigations

Edited by Adrian Linacre

CRC Press
Taylor & Francis Group
Boca Raton   London   New York

CRC Press is an imprint of the
Taylor & Francis Group, an **informa** business

CRC Press
Taylor & Francis Group
6000 Broken Sound Parkway NW, Suite 300
Boca Raton, FL 33487-2742

© 2009 by Taylor & Francis Group, LLC
CRC Press is an imprint of Taylor & Francis Group, an Informa business

Library of Congress Cataloging-in-Publication Data

Forensic science in wildlife investigations / editor, Adrian Linacre.
    p. cm. -- (International forensic science and investigation series)
   Includes bibliographical references and index.
   ISBN 978-0-8493-0410-1 (hardcover : alk. paper)
   1. Wildlife crime investigation. 2. Forensic sciences. I. Linacre, Adrian. II. Title.
III. Series.

   HV8079.W58F67 2009
   363.25'98--dc22   2009000682

Visit the Taylor & Francis Web site at
http://www.taylorandfrancis.com

and the CRC Press Web site at
http://www.crcpress.com

Table of Contents

Series Preface

The modern forensic world is shrinking. Forensic colleagues are no longer just within a laboratory but across the world. E-mails come in from London, Ohio and London, England. Forensic journal articles are read in Peoria, Illinois and Pretoria, South Africa. Mass disasters bring forensic experts together from all over the world.

The modern forensic world is expanding. Forensic scientists travel around the world to attend international meetings. Students graduate from forensic science educational programs in record numbers. Forensic literature—articles, books, and reports—grows in size, complexity, and depth.

Forensic science is a unique mix of science, law, and management. It faces challenges like no other discipline. Legal decisions and new laws force forensic science to adapt methods, change protocols, and develop new sciences. The rigors of research and the vagaries of the nature of evidence create vexing problems with complex answers. Greater demand for forensic services pressures managers to do more with resources that are either inadequate or overwhelming. Forensic science is an exciting, multidisciplinary profession with a nearly unlimited set of challenges to be embraced. The profession is also global in scope—whether a forensic scientist works in Chicago or Shanghai, the same challenges are often encountered.

The International Forensic Science Series is intended to embrace those challenges through innovative books that provide reference, learning, and methods. If forensic science is to stand next to biology, chemistry, physics, geology, and the other natural sciences, its practitioners must be able to articulate the fundamental principles and theories of forensic science and not simply follow procedural steps in manuals. Each book broadens forensic knowledge while deepening our understanding of the application of that knowledge. It is an honor to be the editor of the Taylor & Francis International Forensic Science Series of books. I hope you find the series useful and informative.

Preface

The interest in wildlife crime investigation has increased greatly in the last few years due in part to an increasing awareness of the impact of man on the environment. Ten years ago there were few papers published in mainstream forensic journals on the use of scientific tests for non-human samples. This has been altered profoundly with an ever-increasing number of research groups applying scientific tests to both animal and plant studies. It is against this growth area that this book is designed to cover the developing areas of forensic science that can be applied to wildlife crimes. In part the tests described in the following chapters are applied from their use in other areas of research, predominantly from human-based research. This is primarily a reflection of the lack of research funding made available for wildlife crime investigations. It is interesting to note how crimes against animals and the environment create media interest and public concern, yet the prosecution of these cases is of a low priority, as is the allocation of government funding on research.

The book is aimed at those with a particular interest in forensic wildlife investigation but who are not necessarily active in the area. The scientific material is written at a level such that anyone with an interest in biology will be able to understand the material provided. Basic knowledge that is required is provided in separate boxes to prevent the flow of the chapter being interrupted. Within the chapters there are case examples to illustrate the application of the scientific test. Those involved with the prosecution of wildlife crimes, both from the legal and police professions, will gain from the case examples provided.

The rationale for requesting tests depends on the law that may have been breached. International law covering the cross-border trade in endangered species is subject to regulations through the Convention on the International Trade in Endangered Species of Flora and Fauna (CITES). The role of CITES is explained in chapter 2 by John Sellar, of the CITES Enforcement Division, and sets the scene for the following chapters that cover the applications of scientific tests to answer questions relevant to wildlife crime investigations.

The range of wildlife forensic science tests available depends on the question posed. In wildlife crimes a prime question is 'What species is this?' This is because there is extensive legislation, both international and national, on the trade or illegal killing of defined species. The unambiguous identification of species can be determined by a number of tests depending on the sample type. If there is gross morphology, then visual identification may be possible and sufficient. This might be for eggs of protected birds, turtle shells of protected species or skins from protected big cats or bears. If visual inspection is not possible due to the incomplete nature of the sample, but hairs or feathers are available, then microscopic inspection is an option.

The use of microscopy in forensic science is paramount and a standard tool in species identification. If this simple technique can answer the question asked, then there is no reason to proceed with more expensive tests. The methodology and application of microscopy is described in chapter 3 by members of the Wildlife Institute of India and draws upon extensive experience in this area.

It may be that the type of material is inappropriate for microscopy, and if the species present is required to be identified, then molecular techniques can be employed. DNA testing is now a standard method in the identification of species from trace materials. DNA is also able to link two or more samples to the same individual organism. These linkage types of testing are in many ways similar to the DNA profiling technique used in human identification that is already familiar to the forensic science community. It is entirely relevant that two chapters describe the use of DNA in, firstly, species testing (chapter 4) and then linkage methods, including paternity testing (chapter 5).

A question that is increasingly asked is 'From what part of the world did this sample originate?' In part DNA can answer this question, provided that there is extensive research prior to the test, but the use of naturally occurring isotopes can answer this same question. This is a relatively new field, with the potential application only now becoming realised and utilised; these methods of isotope analysis are described in chapter 6. Development of this type of testing is inevitable in the future, and it may play a much larger role in forensic science in time to come.

The range of species that fall within the remit of wildlife crimes is extensive, ranging from ferns and orchids to great whales. The number of plant species protected by CITES is much greater than animal species, yet mammalian species are those that receive the highest profile. The plight of the tiger, rhino and great panda is reported in the media, yet the extinction of tree species rarely receives attention. Many amphibian and reptilian species are threatened either by loss of their habitat or the trade in live species or their body parts, yet their plight receives little attention. The following chapters focus on animal wildlife crimes, although the reader should be aware that many of the methodologies can be applied equally to nonmammalian animal species and plant species. The case examples used are predominantly from cases involving mammalian species, which may not be a reflection of the actual wildlife crime situation but, rather, of those alleged crimes that are reported and acted upon.

The Editor

Adrian Linacre graduated with an honours degree in zoology from the University of Edinburgh in Scotland in 1984. After this he completed his doctorate studies in molecular neurobiology, using the pond snail as a model system, gaining his D.Phil. from the University of Sussex in 1988.

Appointed as a lecturer in forensic science at the University of Strathclyde in 1994, Dr Linacre's focus of research was on the use of DNA for identification purposes. While undertaking human identification as part of his appointment and working in the criminal justice system, transferring similar methods to non-human samples was an obvious step. In collaboration with other research groups, particularly that headed by Professor Jim Lee at the Central Police University in Taiwan, novel methods in species identification were developed. These included the first DNA test for the identification of cannabis and the first use of DNA to link cannabis samples for the courts. Wildlife testing on rhinos and elephants soon followed, showing how trace material from traditional medicines and ornaments can be isolated and characterised.

Adrian Linacre is currently a senior lecturer at the University of Strathclyde and is on the editorial board of *Forensic Science International: Genetics*. He is a registered forensic practitioner in the areas of body fluid testing, blood pattern analysis and DNA as well as a speciality assessor for the Council for the Registration of Forensic Practitioners (CRFP). He is also Lead Assessor for the newly formed Natural Sciences speciality for the CRFP. A review on the use of low template DNA was commissioned by the UK Home Office, and Adrian Linacre was one of the three authors of the report published in 2008. He is co-author of the book *An Introduction to Forensic Genetics* (2007, Wiley) and has published more than 60 papers in international journals.

His interest in wildlife research continues, and collaborations with the Royal Zoological Society of Scotland, National Museums of Scotland and Wildlife DNA Services are aimed at greater acceptance and awareness of methods for wildlife investigations.

Contributors

S.P. Goyal has been working as a Scientist 'SF' at the Wildlife Institute of India, Dehradoon, India, since 1986. He has been working in wildlife conservation of mammals for more than 25 years. Recently, Dr. Goyal has initiated work on wildlife forensic science to provide assistance to enforcement agencies for implementation of the Wildlife (Protection) Act 1972 (India) and other international obligations. One of the new initiatives under this is to establish hair characteristics of Indian mammals and develop protocols for identifying species from hair for use in wildlife forensics and in determination of food habits of carnivores.

Rob Ogden is project manager of the Wildlife DNA Services unit at Tepnel Research Products and Services, UK, where his work focuses on the development and application of genetic methods to wildlife law enforcement and conservation management.

Since completing his doctorate in population genetics and evolutionary biology, he has undertaken a wide range of applied research projects including the development of DNA profiling systems in bird and mammal species and the production of species identification tests for illegally traded wild meat, traditional medicines and textiles. His current research interests include the genetic identification and traceability of fish and timber products to support both enforcement and certification schemes. He routinely provides forensic analysis and expert witness testimony in UK wildlife crime prosecutions and is a member of the UK wildlife crime forensic working group and the International Society for Forensic Genetics.

Dr. Ogden is also a director of the TRACE Wildlife Forensics Network, an international NGO that promotes the use of forensic science in wildlife law enforcement. Through TRACE, he runs wildlife DNA forensic training courses for participants from across the world, and he provides advice to national and international policymakers. He holds an honorary research position at Bangor University, UK, where he is involved in a number of collaborative wildlife forensic genetic research projects and Ph.D. student supervision.

Vivek Sahajpal attained his master's degree in forensic science from the Punjabi University, Patiala, in 2001. His concern for wildlife conservation brought him to the Wildlife Institute of India, Dehradoon, where an initiative had been taken to develop forensic techniques for dealing with the menace of wildlife crimes in India. Sahajpal has worked on diverse techniques in wild-life forensic science, starting from morphological, microscopic, protein, DNA and other advanced analytical techniques. Trichotaxonomy has been one of his major fields of interest, and he has worked on the hair structures of more than 100 protected Indian mammal species. A reference manual depicting the microscopic hair characteristics for various protected mammals of India has been compiled by Sahajpal and his coworkers and is soon to be published by the Wildlife Institute of India for use by various law enforcement agencies. Sahajpal has several years of experience in dealing with the analysis of wildlife offence case exhibits. He has pursued his Ph.D. on microscopic hair characterization of some highly protected Indian mammal (artiodacyls) species using microscopy, keratin and mitochondrial DNA-based techniques. Sahajpal is also a member of the European Hair Research Society and the Indian Hair Research Society.

John M. Sellar served with the Scottish Police Service for 23 years before moving in 1997 at the request of the UK Government to the United Nations as an enforcement officer of the Secretariat of the Convention on International Trade in Endangered Species of Wild Fauna and Flora (CITES). His role there is to assist countries in combating illegal trade in wildlife, and he acts as the focal point between CITES and international bodies such as Interpol, the World Customs Organization and the UN Office on Drugs and Crime. During his period of work with CITES he has conducted more than 150 missions to 55 countries and territories. Many of these missions have examined species-specific illegal trade in such species and commodities as caviar, ivory, orangutans, rhinoceros, Tibetan antelope and tigers. He is a member of the Forensic Science Society and the International Association of Chiefs of Police.

Shanan S. Tobe obtained his Honours B.Sc. from Laurentian University in Sudbury, Canada, and his M.Sc. and Ph.D. from the Centre for Forensic Science at the University of Strathclyde in Glasgow, UK. His Ph.D. research focused on the identification from mixtures of mammalian species often encountered in forensic investigations. Dr. Tobe has had his research featured in the BBC TV programme 'The Tiger Trail', where it was used to identify traces of tiger bone in Traditional East Asian Medicine (TEAM). He has presented talks at various international conferences, including the International Association of Forensic Sciences, the International Society for Forensic Genetics and the American Academy of Forensic Sciences.

Nature of Wildlife Crimes, Their Investigations and Scientific Processes

1

ADRIAN LINACRE

Contents

1.1 Introduction

The aim of this chapter is to introduce the reader to the variety of wildlife crimes and their investigations. Allied to this is the sample types encountered. Prior to getting into the subsequent chapters on the scientific testing being employed, it is necessary that the tests are validated and fit for purpose. This chapter will cover this most important part of wildlife crime investigations.

1.2 Importance of Wildlife Crimes

If there are no dedicated wildlife crime investigators available, then the priority given to wildlife investigations may be less than that compared to other crimes that require police investigations. National police forces have a limited budget and the priority of investigating wildlife poaching compared to violent crimes such as murders or sexual assaults inevitably will be low. In a few countries there are dedicated wildlife police officers or, for instance, a specialist group at an airport that has a specific task. These types of dedicated staff are few and only employed within a very few countries. If the alleged wildlife crime is investigated by personnel not familiar with wildlife crimes, and the incident is treated with less importance than other incidents that they normally investigate, then it is possible that the wildlife crime scene samples are not recorded and packaged to the highest standards. This can have implications for the future investigation and any trial.

1.3 Sample Types

The types of samples encountered in wildlife crime investigations will be highly varied as the samples may be from amphibians to mammals, wood fragments to flowers, and from complete skins to powders and oils. If such samples are received into the forensic science laboratory, then the tests that could be employed are dependent upon the allegation.

1.4 Case Assessment

Prior to embarking upon any examination it is necessary to have background information. This will not only be the allegation, i.e., which legislation is alleged to have been transgressed, but also as much additional information as possible. This will include from where the sample was collected, when it was collected and, if there is a defendant, any alternative explanations. Forensic science, including wildlife forensic science, should be conducted within a framework of a case. This means that the scientist asked to conduct the examination is fully aware of the nature of the allegation and therefore which examinations to perform that will either support the allegation or will help support a counterproposition. The examinations that can be conducted also depend upon the capability of the scientist within the laboratory and the capability of the laboratory. If the laboratory does not have DNA equipment, then there are limitations in the testings that can be performed. Equally, if the laboratory is not staffed by someone competent in microscopy, then, even if the equipment is present, the examination cannot be conducted. It would be valuable if, when a laboratory is unable to perform a particular test and there is a real need to conduct the test, there is knowledge of where to find someone who can. These organisations are mentioned in section 1.7 towards the end of this chapter.

1.4.1 Example of Case Assessment (1)

The body of a deer is found in a field with the antlers removed. The deer is alleged to be from a protected species and the killing of the animal species is contrary to national legislation. The antlers are a valuable commodity. The investigator should consider the following questions:

- An obvious question is, did the animal die of natural causes or was it killed illegally?
- The second question is, what species is it?
- The third question would be, where are the antlers? And, if any antlers are recovered, did they come from this particular deer?

If possible, a vet should be called to determine how the animal was killed. If the animal died of natural causes, then the only possible transgression is the collection of the antlers (if this is an offence). If the animal was killed illegally, then there will be reason to continue the investigation.

It is unlikely that the entire carcase will be transported back to the laboratory, but samples that should be collected will include hair and blood samples. Of potential relevance will be the base of the antlers, as marks on the body may assist in future identification of a tool if found. A full photographic record of the carcase should be made. The decisions as to what should be recorded and collected need to be made by someone competent in evidence collection. If the evidence is compromised at the scene by poor collection, then any future investigation may be undermined.

Depending upon the deer species present, microscopy of the hair may be sufficient to identify unambiguously to species level (discussed in chapter 3). For many species of deer this may not be possible, in which case the blood (or the hair) can be used to determine the species using DNA typing (discussed in chapter 4).

If the antlers are found, then linkage between the antlers and the carcase of the deer can be performed by DNA profiling (discussed in chapter 5). This is dependent upon the quality of the antlers.

1.4.2 Example of Case Assessment (2)

A statue suspected of coming from ivory is intercepted at an airport.

- The first question is, is the statue made from ivory?
- If it is made from ivory, then from what species did it originate?
- If from a species listed on one of the three appendices of the Convention on International Trade in Endangered Species of Wild Fauna and Flora (CITES), could the ivory be exempt due to the age of the sample?

The identification of ivory can be performed by microscopy or by elemental analysis. The application of microscopy depends upon the amount of material present and whether the structure of the original ivory has been altered to prevent the identification. A range of methods using elemental analysis is also possible. There are methods for the isolation of trace amounts of DNA from ivory, including from ivory statues, in which case species identification is possible. Possession of certain ivory samples is permitted prior to a particular date. For instance, the European Union permits the trading of genuine antiques if the samples are older than 1 June 1947. At present there are few methods that can accurately differentiate ivory prior to 1947 and post-1947. One specialist technique that can be used is the examination of the types of radioisotopes released by nuclear weapons, as these would only be in the environment from 1945 onwards.

1.5   Validation

The term *validation* can be defined as the process by which a test is determined to be fit for purpose, reliable, can be defended if challenged, and is accepted by the wider scientific community. Any new scientific test must be validated prior to its introduction to forensic science. Forensic science ultimately is part of the criminal justice system, and this is the case as much for wildlife crimes as for crimes against humans. It is necessary that methods used in wildlife investigations are validated to the same standards as those used in crimes against humans. The process of validation is described clearly by the Scientific Working Group on DNA Methods (SWGDAM). The steps presented in the following section are illustrative of those required to validate a procedure used in wildlife investigations. It may be that not all are relevant to every procedure but provide an outline of the process required. The experiments conducted to demonstrate the validity of the test should be documented with a comment on the outcomes of the validation studies.

1.5.1   Characterisation of Type of Sample

It is necessary that the procedure to describe the item is documented fully. This could be its general appearance and dimensions. This will vary depending upon the type of sample.

1.5.2   Specificity

The test should only detect the substrate, object or species for which it is designed. For instance, the microscopy of hair may identify a particular species of bear if a series of morphological characteristics is present. The unambiguous identification of the species needs to be determined using the procedure adopted, and the test needs to identify this particular species and no other. It will be necessary to demonstrate that all closely related species are not misidentified.

1.5.3   Sensitivity

Depending upon the test, it will be necessary to document the lowest amount of the sample by which the test will generate a result. In terms of DNA typing, it is possible to make serial dilutions of a known amount of starting material and record the concentration at which the DNA can be detected and below which no result should be expected.

This step may not be applicable in microscopy as a hair is either present or not; however, the number of hairs that are required to be examined to be confident of identification may be recorded.

1.5.4   Stability

A record of the environmental factors that will affect the results should be provided. In particular, a record should be made if any environmental factor could lead to an erroneous result rather than no result. Environmental factors include heat, humidity, sunlight and chemical treatment.

1.5.5   Reproducibility

A key set of experiments is reproducibility studies. Reproducibility will document both the precision and the accuracy of the test. It should be noted that a method may generate precise data, but not accurate data. An example in human identification is that all the sizes of DNA fragments examined are one DNA base bigger than they should be, but as all DNA fragments are inaccurate by one DNA base, a precise result can be recorded. The number of samples required to demonstrate reproducibility depends upon the data obtained, but the data should be subjected to standard statistical tests.

1.5.6   Questioned and Known

A laboratory may receive both questioned (case-related) and known (reference) material, and a procedure to minimise the opportunity for contamination of the case-related sample by the reference sample should be documented. Laboratories working with trace levels of material are particularly aware of this problem and will examine questioned and reference samples either in totally separate laboratories or by using different staff. It is important that control samples such as negative controls are used to monitor the opportunity for contamination.

1.5.7 Peer Review

Although not essential for validation, a procedure would gain greater acceptance by the external community if it were published in a peer reviewed journal. There are a number of international science journals that will accept validation studies (*Journal of Forensic Sciences* and *Forensic Science International* are two such journals). Organisations that develop a new method are encouraged to publish the data in such an international journal.

1.5.8 Validation Example

The following describes the steps that might be undertaken in developing a novel test to indicate the genetic relationship between members of the same species of bird.

I. A statement is required to define the aim of the test. In this case it can be stated that the aim is to establish whether two birds of the same species could be genetically related and, if this could be the case, then whether a probability of genetic relatedness can be determined.

II. A description of the sample received, including from whom and when, is needed. If this is to be a reference sample for the test, then its species origin must be vouched for as there is no point in using material for this type of test that is of questionable origin. The description of the actual item may include its weight and dimensions.

III. The DNA should detect specific regions (called loci, with the singular being a locus) within the entire DNA (the genome) of this particular bird species. A series of specificity tests is required to ensure that the DNA test will only identify the particular species to be tested and that the test will only identify the particular locus in the entire genome of the species. These studies (not insignificant) need to be documented. For these studies it will be needed to obtain voucher specimens from related species to ensure no cross-reactivity. A range of completely unrelated species, including human, should also be tested to confirm that there is no cross-reactivity with DNA from species that are more commonly encountered.

IV. The test should be conducted on samples taken from a range of material collected from the bird, including blood and feathers, to ensure that the same result is obtained regardless of the type of sample taken from the same bird.

V. Serial dilutions of the DNA can be made to determine the concentration of DNA at which a result can be obtained. There will be a threshold below which a result may not be obtained. The reason for this is that exceptionally low amounts of DNA, such as in trace contamination, may not be detected, and only if known amounts of DNA are present will a result be obtained.

VI. Leaving blood and feather samples for defined periods of time, typically ranging from 1 day to 3 months, in a number of different conditions, such as variable heat and humidity, will demonstrate the effect of environmental conditions. Under certain conditions a positive result may be expected, but under other conditions no result will be obtained. It needs to be demonstrated that the conditions will produce either a positive result or no result and that misidentification is not possible.

VII. After the test has been subjected to initial testing, it is necessary to undertake reproducibility trials on a range of reference samples from the same species to ensure that positive results are obtained. If the same sample is tested 10 times, it must generate exactly the same result.

VIII. As the aim is to provide evidence of a genetic link or otherwise, it is necessary to show that the test will show inheritance from one generation to the next. This can only be undertaken with known breeding pairs and their offspring. Additionally, it is necessary to know that unrelated birds of the same species have variable DNA types as an indication of the polymorphic nature of the DNA locus. A population survey may need to include up to 100 birds of the same species to generate a meaningful database of variable (allele) types.

IX. Based upon all the previous steps, a standard operational procedure (SOP) can be written. This SOP document should cover all of the previous steps, and within the file there should be the experimental detail undertaken.

X. Finally, it may be that the test is sufficiently novel that the methodology can be published in a peer reviewed journal.

It should be noted that this description of a validation procedure is brief and gives only the outline of the steps required. Readers are recommended to view the SWGDAM guidelines at http://www.bioforensics.com/conference04/TWGDAM/SWGDAM.pdf.

1.6   Presentation of Evidence at Court

A final arbiter of the validity of the test will be the court. This is dependent upon the jurisdiction of the country where the case occurred. Additionally, the role of the forensic scientist may vary from country to country. Typically a written report will be produced setting out the examination conducted and the results obtained. It may be that the written report is sufficient; however, presenting evidence verbally in the court is standard in many countries and would be the same in wildlife crimes. It is essential that the evidence given is in the framework of the allegation provided and an alternative scenario (if provided) by the defence is considered. In chapter 5 the reader is introduced to the concept of a likelihood ratio, where the scientist considers the probability of finding the result if the allegation is true compared to the probability of obtaining this same result if the defence position is true. In some types of evidence, such as DNA, it is possible to provide a number, but this is rarely the case, and instead the scientist states whether the evidence supports one of the scenarios more than the other, and whether the evidence is strongly in favour of the scenario.

1.7   Forensic Science Organisations Involved in the Investigation of Wildlife Crimes

There are a number of international organisations that can provide either information through their Web sites or more specific help by actually helping with the analysis.

The World Wide Fund for Nature was founded in 1961 and remains the leading international organisation to highlight and raise funds for the protection of wildlife throughout the world. Due in part to its wide-ranging work, the organisation goes under the name of the WWF. More information can be found at www.wwf.org, where there are links to all its other Web sites.

The Convention on the International Trade in Endangered Species of Flora and Fauna (also known as CITES; www.CITES.org) provides on its Web site a valuable source of information on the role of CITES and the species that are protected under international legislation. The work of the enforcement of CITES legislation is discussed in chapter 2 of this book.

TRAFFIC is a monitoring network working to ensure that trade in wild plants and animals is not a threat to the conservation of nature. As such, it is linked with the work of the WWF and CITES. Information about the role of TRAFFIC can be found at www.traffic.org.

The World Society for the Protection of Animals (WSPA) is primarily involved with the protection of animals from cruelty, but this can inevitably transgress national legislation. WSPA has supported the development of scientific tests to aid in the investigation of cruelty and the use of protected species such as bear species used in the production of bear bile. More information can be found at www.wspa-international.org.

The International Fund for Animal Welfare (IFAW) was established with similar aims to that of WSPA and is very much focused on the protection of animals in wild populations that are affected by the activity of man. More information can be found at www.ifaw.org.

The U.S. Fish and Wildlife Service, with a laboratory in Oregon, has been a main organisation aiding in investigations of wildlife crimes. Many tests that are now used in wildlife science have been developed by this organisation. Full details of the activities of the U.S. Fish and Wildlife Service can be found at www.fws.gov.

TRACE is an organisation that aims to provide integrated methods in the investigations of wildlife crimes and acts as a reservoir of useful information. TRACE can provide advice with requests relating to matters relating to wildlife crimes. TRACE can be found at www.tracenetwork.org.

Illegal Trade and the Convention on International Trade in Endangered Species of Wild Fauna and Flora (CITES)

2

JOHN M. SELLAR

Contents

2.1 Background

Since time immemorial, that most dangerous of species, *Homo sapiens*, has used animals and plants to his own ends. He has fed upon them, clothed himself with their skins, and treated himself with their medicinal properties. Throughout the centuries of man's existence, this exploitation of wild fauna and flora has, in essence, probably changed little.

It was not until perhaps the 1800s that man began to reflect on the ways that he used the species with which he shares Earth. Even then, many of the first pieces of legislation adopted by "developed" nations tended to concentrate on animal welfare issues and were designed simply to punish acts of cruelty. Initial international discussion on what we might now regard as conservation often centred on the colonial powers' anxieties that the hunting of "big game" in African range states was threatened by overexploitation. Many people would argue that, even today, we still place too much emphasis on what has come to be known as "megafauna," that is, elephants, rhinoceros, large cats, etc.

Whilst a number of international treaties were drafted, and some even ratified, most environmentalists agree that it was with the signing of a draft convention by 21 countries, in Washington, DC, on 3 March 1973 that the first effective steps in regulating wildlife use truly began at a global level. Although still known in some parts of the world as the Washington Convention, what entered into force on 1 July 1975 is more properly called the Convention on International Trade in Endangered Species of Wild Fauna and Flora. It is commonly known by its acronym of CITES.

CITES is widely regarded as one of the most successful of all conservation treaties. The mere fact that it now has 172 signatory states (known as parties) illustrates the manner in which it has grown and continues to be viewed as relevant. "Producer" countries appreciate that the import controls of CITES, as well as control at the place of export, offer support to their efforts to combat exploitation of their natural resources by poachers and illicit traders. On the other hand, "consuming" nations welcome the controls that enable their legitimate dealers to obtain supplies at what should be sustainable levels.

There are a number of misconceptions about CITES. Not the least of these, and widespread amongst law enforcement officers, is that a principle aim of CITES is to ban wildlife trade. Whilst it is certainly true that the Convention recognizes, and seeks to address, the dangers of uncontrolled trade, it should rather be viewed as a regulating mechanism for trade. Indeed, it has been noted at meetings of the Convention that "commercial trade may be beneficial to the conservation of species and ecosystems and/or to the development of local people when carried out at levels that are not detrimental to the survival of the species in question."

A vital element of the Convention is its three appendices. These list the species controlled by CITES and determine the level of control, as follows:

- Species threatened with extinction that are or could be affected by trade (Appendix I)
- Species not necessarily in danger of extinction but which could become so if trade in them were not strictly regulated (Appendix II)
- Species that individual parties to the Convention choose to make subject to regulations and for which the cooperation of other parties is required in controlling trade (Appendix III)

The Convention is implemented by national management and scientific authorities, which are assisted by the CITES Secretariat. It is based in Geneva, Switzerland, and is staffed by a small group of individuals with a range of wildlife-, law-, and trade-related experience. At present, only one staff member is devoted to illegal trade issues. The Secretariat maintains a Web site (www.cites.org) that provides detailed information regarding the workings of the Convention. The Web site also contains a directory of national CITES authorities, including law enforcement authorities in many countries.

Several forms of crime are referred to as being the second largest in the world behind drugs, and this is sometimes said of wildlife crimes. The CITES Secretariat makes no such claim. What is not in question, however, is that unregulated or illegal wildlife trade is the second greatest threat to many endangered species in the world, next to habitat destruction.

Illegal trade in wildlife can be big business, and there are many examples of this.

During 2001, the Secretariat examined the trade in caviar leaving one Middle East country. During a 10-month period, this amounted to a wholesale value of more than USD 20 million. The majority of this was determined to have originated from poaching activities in the Caspian Sea and had been "laundered" using genuine CITES re-export documents that were obtained by fraud, threats of violence and corruption of officials.

One "shahtoosh" shawl, made from the wool of a Tibetan antelope (*Pantholops hodgsonii*), can retail at more than USD 20,000.

2.2 How Is Illegal International Trade Conducted?

Smuggling

- Across a border without a customs or control point
- Across a border by hiding the specimens:
  - In luggage
  - Under clothes
  - Inside vehicles
  - Using boats and planes
  - Inside containers
  - In crates containing dangerous animals (allegedly or not)
- By post (including eggs, parrots, birds of prey, stuffed animals and their skins, live reptiles, ivory, medicines, and plants)
- By changing the appearance of the specimen

Fraud

- By false declarations
- By bribing officials
- By altering or modifying genuine CITES permits and certificates
- By forging permits, certificates, security stamps, and authorizing officers' signatures

In other words, the illicit trade takes forms that would be immediately recognizable to customs officers as similar or identical to those used for a wide range of other contraband, such as narcotics.

2.3   Is Organized Crime Involved?

This often-posed question cannot be answered by referring to convictions of leading members of the "Mafia," but the CITES Secretariat and UN Office on Drugs and Crime have researched this issue. What is very clear is that there are many indicators that demonstrate that organized crime groups or criminal networks are involved in, or linked to, wildlife crimes. These include:

- Organized structure to poaching, including the use of gangs and supplying vehicles, weapons, and ammunition
- Exploitation of local communities
- Provision of high-quality lawyers
- Corruption of judicial process
- Violence towards law enforcement personnel
- Corruption of law enforcement personnel
- Exploitation of civil unrest
- Financial investment in "start-up" and the technology needed for processing and marketing
- "Inviolability" displayed by those involved
- Sophistication of smuggling techniques and routes
- Use of "mules" and couriers in cross-border smuggling
- Payments to organized crime groups
- Use of persons of high political or social status
- Sophisticated forgery and counterfeiting of documents
- Use of sexual bribes or blackmail and other corruptions of officials
- Use of fake or "front" companies
- Fraudulent advertising of wildlife and widespread use of the Internet
- Previous convictions for other types of crimes
- Organized crime group members' use or ownership of wildlife
- Huge profits

2.4   The Importance of Forensic Science

Wildlife crimes are essentially no different from any other form of criminality, and the full range of forensic science expertise and support can potentially be bought to bear from one end of the illicit trade chain to the other. The following are just a few examples of the forms such support may take:

- Ballistics—To link bullets recovered from carcasses, or cartridge cases found at a poaching scene, with firearms subsequently seized from suspects. Similarly, "matching" bullets found at different poaching loci can demonstrate the involvement of repeat offenders and cross-border poaching.
- DNA profiling—Often essential in accurately identifying species parts and derivatives, especially when one is examining the ingredients of a final "product," such as caviar or traditional medicines. Such profiling is also increasingly used in helping to reveal the geographical origin of a specimen.
- Isotopic profiling—Another technique that has great potential for geographical origin determination.
- Latent fingerprint impressions—Can be useful for obvious instances, such as who handled a weapon, but also of great potential in identifying who handled the packaging used for wildlife smuggling or who handled a fake CITES document.
- Morphology—Given access to appropriate reference collections, this is often the first step in determining what species one is dealing with. This can involve whole specimens, right down to fur and feathers.
- Pathology—A wildlife law enforcement officer investigating the suspected "murder" of an animal needs to know how it died, just as much as his counterpart in human homicides. Therefore, the assistance of forensic veterinary surgeons, experienced in necropsy, is essential.
- Questioned document examination—Can not only determine whether a CITES document is genuine or counterfeit but may also help prove who forged it.
- Toxicology—Knowing what poisoned a rare raptor and subsequently finding the same chemical in a suspect's possession is important evidence.

2.5 Why Isn't Forensic Science Used More?

Given its very considerable potential, this is a valid and important question.

The CITES Secretariat has, for many years, had a Memorandum of Understanding with the Clark R. Bavin National Fish and Wildlife Forensics Laboratory (www.lab.fws.gov). Located in Oregon and operated by the U.S. Fish and Wildlife Service, it is one of the very few laboratories in the world that is devoted solely to wildlife crimes. The laboratory has agreed to offer its services free to any party to CITES. Examinations will be conducted without charge, but the costs of the attendance of scientists to give evidence in court may need to be recompensed.

The CITES Secretariat also agreed with the laboratory that it would act as an international repository for ballistic evidence and that parties could submit bullets, cartridge cases, etc., so that cross-border poaching and repeat offending could be easily and rapidly identified.

As it happens, relatively few parties have made use of the general offer of assistance, and hardly any have submitted evidence to the ballistic repository. It seems that there may be a variety of explanations for this.

One probably has to first understand wildlife law enforcement around the world. Historically, especially in those countries that were previously colonies of western nations, wildlife was seen as the property of the rulers, the aristocracy, the military and forestry officer cadres, and the political elite. Wildlife, particularly animals that were targets of sport hunting, was reserved for them. In those days there was little need to deploy well-equipped or intensively trained enforcement staff. Game scouts and forest guards in the savannah of Africa and forests of Asia carried sticks (almost as much as a symbol of office as a weapon) and patrolled "beats," just as a police constable would then have done. These officials were there to deter and detect those people who might, from time to time, forget their place in society and seek to take their share of nature's bounty, be that an animal or wood for their fire. They were there to deal with the nuisance element that might spoil a day's shooting or reduce the number of possible hunting trophies. Alongside their enforcement duties, such staff were heavily engaged in recording population numbers, maintaining roads and tracks, helping control timber harvesting and planting, watching for fires, and fighting those fires when they occurred.

It's a sorry reflection of the priority given to wildlife law enforcement that a similar situation can be found in many developing countries today, decades, if not centuries, after such enforcement regimes were first established. The majority of wildlife law enforcement in the habitats of endangered species around the world is based upon anti-poaching activities. Consequently, those tasked with enforcement seldom have the "policing" skills necessary to respond to and combat today's wildlife criminals. They have not received training in interview techniques, intelligence-gathering, giving evidence in court, or the myriad of other basic skills that a police officer must have. Poachers, smugglers, and dealers are likely to be better armed, better equipped, better educated, better paid, and better organized than many wildlife law enforcement officers. Also, because the focus is on anti-poaching work, the majority of enforcers are in the field and few are deployed in the towns, cities, and ports where the trade and smuggling occur.

Times are changing and improvements are taking place. However, forensic science support, or even awareness of it, remains very low on the list of priorities in many parts of the world. If the man or woman at the "sharp end" doesn't appreciate the significance of what could be important evidence, he or she is unlikely to collect it, or label and record it in a manner that will be acceptable in court and then send it to an appropriate laboratory or expert.

One also has to consider the situation in many of the countries where illegal harvesting of fauna and flora is taking place. The police there may not have access to forensic support for murders. No scene-of-crime officers attend burglaries in the cities there. Imagine, then, the reaction that a game scout, perhaps one who has attended a CITES training course, is likely to encounter if he returns from the field with his pack full of labeled cartridges and plaster casts of vehicle tracks. If he has knowledgeable and understanding bosses, he might be praised. But then they'll have to identify a laboratory to which they can submit the evidence, and they'll have to find the money to get it there. However, he also probably risks being ridiculed by other (less aware) officers or counterparts in other agencies who will ask, "You want to do what?"

2.6   The Future

Given the above, it is little wonder that one encounters interested and committed, but frustrated, forensic scientists who have devoted hours to developing or adapting techniques for wildlife crimes only to find that hardly anyone seeks out their assistance.

Lack of awareness of forensic science, scientists, and laboratories can be tackled. It probably cannot be addressed quickly, but we'll get there gradually. The major problem seems to be expense. That can range from simply finding the money to send "evidence" from one continent to an appropriate forensic facility on another, to being able to afford the fees that some facilities charge. But, of course, it is also expensive and time consuming for facilities to develop the profiles or reference collections needed for comparative analyses.

Cost isn't just a factor for the developing world, though. Not so many years ago, customs and police in the United Kingdom, for example, thought nothing of submitting potential evidence for forensic examination. This was because there would be an in-force or regional forensic laboratory and the expense was absorbed by the agency's budget. Increasingly, however, forensic services have become independent providers who need to raise their own operating revenues. Consequently, a detective may have to consider much more carefully what should be examined. If this is a consideration in "the West," consider the situation elsewhere.

The developed world must be willing to support law enforcement colleagues elsewhere, and there is considerable need for the provision of training and awareness-raising as well as the subsidizing of forensic examinations. But we'll have to bear in mind the Oregon laboratory's experience, and one suspects that its no-fee offer may be seen by some as simply too good to be true.

From the CITES Secretariat's perspective, we also need greater professionalism from the ground up, so that the law enforcers know what to do and the scientists are appropriately experienced and qualified. The Secretariat has seen many examples of results from researchers, undoubtedly interesting and promising, but where the samples were not collected in an appropriate fashion, no chain-of-custody was maintained, the laboratory did not have the relevant accreditation, and, thus, everything would likely be inadmissible. In their enthusiasm to seek out and provide results, people can fail to distinguish between science and forensic science.

The Secretariat would also welcome greater communication, coordination, and collaboration across the field of wildlife forensic science. There certainly seem to be examples of unnecessary duplication of research taking place, and, in some cases, research time is being devoted to species that seldom enter into illegal trade whilst seriously threatened others might be ignored or receive insufficient attention. Initiatives such as TRACE (the wildlife forensics network) can help promote the above three "Cs" whilst addressing and countering problems.

We don't need a wildlife forensic laboratory in every country. Indeed, we may not need one in each region or on every continent. But we definitely need accessible and accredited centers of excellence, and these seem to be appearing more and more. The wildlife forensic science community certainly appears to wish to work together, and we need to find mechanisms and fora to enable it to do this better.

Foundations, charities, and non-governmental organizations raise and distribute millions of dollars each year for wildlife conservation purposes. Some of this money deserves to be diverted to the provision of forensic science support. As the wildlife forensic science community becomes better coordinated, perhaps through the establishment of some association or formal network, it may be easier to access the funding support that will be needed for a variety of activities.

Equally importantly, the wildlife forensic science community of the future should have greater opportunities to influence political decision-makers and senior law enforcement management so that wildlife crimes receive the attention and priority they deserve. Then, one day, no game scout, forest guard, warden, ranger, or fishery protection officer will hesitate to collect, bag, label, and submit evidence from a poaching scene, smuggling case, or raid on a wildlife dealer's store. The CITES Secretariat eagerly looks forward to that day.

Microscopic Examinations in Wildlife Investigations

3

VIVEK SAHAJPAL AND S. P. GOYAL

Contents

3.1   Introduction

Microscopy is one of the most basic tools required for wildlife forensic investigations.

The simplicity, cost, and time effectiveness of microscopy make it an apt tool to carry out many forensic wildlife investigations. It can be applied to the examination and comparison of a wide variety of articles like hairs, feathers, claws, and beaks. These sample types are examined routinely in wildlife offenses. Examination of exhibits under the microscope will always provide some valuable information and investigative leads in a nondestructive manner. Hence, for a range of samples microscopic examination is the first step, and can be the only technique required for some species.

There is a wide spectrum of applications of microscopy in wildlife forensic science, and it is applied routinely for species characterizations from hairs based upon microscopic hair characteristics. Since a large proportion of wildlife crime is centered on illegal poaching and the trade of mammalian species, hairs are very frequently obtained as physical evidence. Hence, microscopic examination of hair evidence for microscopic hair characteristics emerges as a powerful method for species characterization and identification. In this chapter we focus more on the types of microscopes required, terminology, methods, and features to be assessed for such examinations.

3.2   Microscopes Used in Wildlife Forensic Science

There is a wide variety of microscopes that can be used in wildlife forensic examinations, but the two types of light microscopes that are essential tools are stereoscopic microscopes and comparison microscopes. A scanning electron microscope (SEM) can be used if available, but due to the costs of an SEM, it is not available in many wildlife forensic science laboratories.

3.2.1   Stereomicroscope

A stereomicroscope is actually comprised of two microscopes that focus on the same point from slightly different angles, producing a three-dimensional image of the object being viewed. As opposed to compound microscopes, the image produced with a stereomicroscope is always upright. This type of microscope is used to view objects under low magnifications, below 100× magnification. The working distance (between platform and objective lens) is much longer than with a typical compound microscope, and as such the surface structures of large objects can be viewed with great ease. This type of microscope can work in reflected as well as transmitted light mode, depending on the types of samples and needs of the examiner.

A stereomicroscope can be used for virtually any type of evidence (hair, bills, claws, shells, feathers, ivory, etc.) to study the surface morphology at low magnification. As wildlife forensic science deals with a wide variety of artifacts, stereomicroscopic examination is very useful in preliminary examinations to provide an indication about the species of origin of the exhibits. As such, it is a useful screening method that can aid in narrowing down the potential species from which the sample originated.

3.2.2 Comparison Microscope

A comparison microscope is also comprised of two compound microscopes that are joined together by a bridge, which produces a side-by-side image of the two samples being compared. The two samples being examined may be an unknown sample from a crime scene and a reference sample of known origin. This type of microscope allows for the comparison of surface morphology, diameter, and internal structure of the questioned sample to a reference sample. A comparison microscope is well suited in the study of microscopic hair characteristics as it provides the best method to compare a questioned hair(s) with a reference hair sample where there is a need to obtain more than 100× magnification.

3.2.3 Scanning Electron Microscope (SEM)

SEMs work on the principle of viewing samples with the help of a beam of electrons that scan a sample in an array of picture points. This is useful where a very high resolution is required. SEM is of great use in the study of fur and wool hair for species characterizations. However, it is not obligatory to have an SEM for wildlife forensic studies on hairs.

3.3 Studies on Species Characterizations from Microscopic Hair Characteristics

Species identifications based on microscopic hair characteristics have been used widely in biological sciences for studying food habits, prey–predator relationships, and mammals inhabiting a den or a tree (Mathiak, 1938; Mayer, 1952). The early pioneering work on species characterizations from hair includes studies by McCurtie (1886), Hausman (1920, 1924, 1930, 1932, 1944), and Hardy and Plitt (1940). Descriptive guides on microscopic hair characteristics for some important mammalian species of particular regions have been contributed by Brunner and Coman (1974), Moore et al. (1974), Appleyard (1978), and Teerink (1991).

Species characterization and identification deal with several aspects of the hair, including hair profile, hair type, cuticular characteristics, medulla characteristics, cross-sections, pigment type and its distribution, etc. All these aspects for species characterizations from hair using microscopic methods are discussed in detail in the following sections.

3.4   Hair Profile

The general shape or profile of the hair is of appreciable value in character-izing hairs of various species and is the starting point in the characterization of an unknown sample of hair. The hair can be divided into the root and the shaft. Most mammal species have guard hairs, which have a flattening (shield) toward the distal (away from the skin) end (see figure 3.3a later). Similarly, certain species have guard or fur hairs with a zig-zag appearance. These gen-eral shapes or profiles should be noted carefully while preparing a reference checklist and while examining unknown samples as they help in narrowing down the investigation to the most probable family, genera, or species.

3.5   Types of Hair

The complete covering of hair over a mammal is termed a pelage. For many mammalian species there is a wide range of hair types that cover the body, each serving a particular function such as heat conservation, protection, or as sensory hairs. All these types are the pelage.

Prior to proceeding with a microscopic examination of hair for species identification it is most essential to have a good knowledge about the dif-ferent "types" of hair. Classifications of hair "types" have been provided by many workers (Brunner and Coman, 1974; Moore et al., 1974; Teerink, 1991). Hausman (1920) described only two types of hair, i.e., under-fur and over-hair, a refinement of which was reported by Dreyer (1966). The following five "types" of hair have been recognized (Brunner and Coman, 1974).

Figure 3.1 Cross-section of Tiger (*Panthera tigris*) and Leopard (*Panthera pardus*) vibrissae.

3.5.1 Vibrissae or Whiskers

These hairs are also referred as sensory, tactile, or sinus hairs. Their tactile sensory function is well established. Vibrissae commonly grow around the nostrils, as seen in the felid (cat) species. Vibrissae are usually thicker and stiffer than other types of hair. They are thickest at the proximal (close to the skin) end and gradually taper toward the distal (away from the skin) end. Not much value in species identification has been associated with vibrissae due to a lack of significant variation in structure among mammalian species (Brunner and Coman, 1974). However, we have found some variations in cross-sections of tiger and leopard vibrissae (figure 3.1). This may be of significance in distinguishing between species.

3.5.2 Bristle Hair

These are found in cases involving domestic pigs and wild pigs or boar (family: Suidae). The pelage comprises thick and stout hairs with a flagged or forked tip (figure 3.2a). The medulla is either absent, narrow, or intruding at certain places. The variation in diameter along the length of the hair is minimal. The cross-sections are predominantly oval and circular, but oblong cross-sections have also been observed. The forking of the tip is a useful indicator for the family Suidae. Bristles are predominantly used in some shaving and painting brushes.

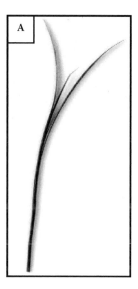

Figure 3.2 (a) Bristle hair found in domestic pigs and wild boars (*Sus sp.*).

Figure 3.2 (b) Various types of hair in a pelage.

3.5.3 Over-Hairs

These hairs are distinct from the main pelage as they are distinctly longer than the rest of the hairs (Brunner and Coman, 1974). These hairs are scattered sparsely across the main pelage and are stiff and straight with elongated tips. The cross-section of over-hair is mostly circular in outline and has little value in species identification unless the cross-section is not circular; this form is found in only a few species (Brunner and Coman, 1974).

3.5.4   Guard Hairs

In most mammalian species the guard hairs form the bulk of the pelage and have the highest significance in species identification. They are long and coarse in appearance and show interspecies variations in their microscopic hair characteristics (figure 3.2b). Guard hairs in many species show the formation of shields (flattening) toward the distal (away from the skin) ends.

Guard hairs may be characterized as primary guard hairs or as secondary guard hairs, depending on their size. Primary guard hairs are the longer of the two types and are more important in species identification based on microscopic hair characteristics.

3.5.5   Under-Hairs

These are short and very fine hairs found on the pelage. They are shorter than guard hairs and can be seen when guard hairs are moved aside or removed (figure 3.2b). They usually show very little variation in thickness from the proximal to the distal ends. Little diagnostic value has been associated with them for the purpose of species identification (Brunner and Coman, 1974); however, recent studies indicate that species can be identified from under-hairs (Rollins and Hall, 1999; Phan et al., 2000).

Wool is an example of an under-hair from mammal species. Shahtoosh wool (under-hair) obtained by slaughtering the Tibetan antelope (*Pantholops hodgsonii*) is frequently obtained as wildlife offense case exhibits, and in such cases the species has to be characterized from the cuticular structure of the under-hair or wool and can be easily distinguished from Cashmere/Pashmina wool (figure 3.3b and c). The Tibetan antelope is both highly prized for the fleece and highly endangered. The species is listed in CITES appendix 1 and therefore trade in Shahtoosh is in contravention of CITES agreements (see chapter 2).

3.6   Hair Structure

Prior to proceeding with microscopic studies of hair, it is imperative to understand the basic structure of a hair. Hair is made of bundles of fibrils that are made of the chemically inert protein called keratin. Keratin is a protein that is also present as rhino horns, fingernails, and velvet of antlers and is composed of amino acid chains that are rich in the amino acid cysteine. Two cysteine molecules have the property of being able to bond together, resulting in the long chains binding to each other and giving much greater strength compared to free single chains of amino acids.

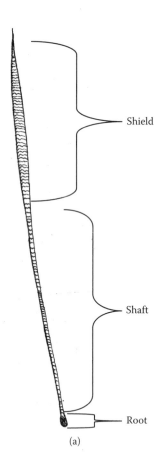

(a)

Figure 3.3 (a) Hair gross external appearance.

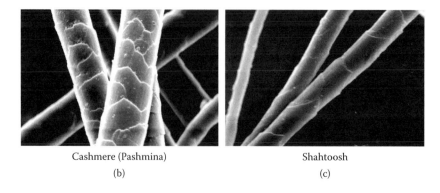

Cashmere (Pashmina)   Shahtoosh
(b)   (c)

Figure 3.3 (b, c) Cashmere and Shahtoosh wool (SEM image).

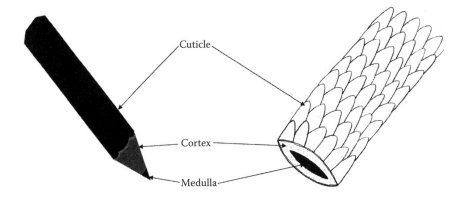

Figure 3.4 Hair structure.

A typical mammal hair is comprised of three layers: the innermost core region or the medulla, an intermediate layer of cortex surrounding the medulla, and the outermost layer of overlapping and transparent scales known as the cuticle (figure 3.4). Perhaps the best way to explain the structure of hair is to compare it with a lead or graphite pencil, where the innermost lead or graphite layer represents the medulla, the wooden portion is the cortex, and the outermost paint layer is the thin layer of cuticle (figure 3.4).

The shape, arrangement, and size of these three layers, viz. medulla, cortex, and cuticle, are the foundations in hair characterization and identification (Brunner and Coman, 1974).

3.6.1 Cuticle Structure

The cuticle, or the outermost layer, comprises a large number of mostly transparent and overlapping scales. Scales may be pigmented in certain cases, such as bat species (*Chiroptera* spp.) (Benedict, 1957). The distal part of each scale usually covers the proximal part of the next scale, resulting in the hair having less resistance from the base to the tip than in the opposite direction (Teerink, 1991). This property can be used to quickly identify which end is the root or the tip without using a lens or microscope. This can be done by taking the hair between the thumb and forefinger and pulling it. The orientation in which it gets pulled out with the least resistance helps to identify the root or tip. The root end is toward the pulling fingers or hand when the hair is pulled out with least resistance.

The shape, size, margin, and arrangement of the scales along the hair shaft vary across species and are used for species characterization and identification. However, the shape, size, margin, and arrangement of scales also vary along a single hair shaft; i.e., the proximal, medial, and distal regions of the hair show differences in their scale patterns.

There are two measurements that can be made based on the scale pattern: the **scale index** (Hausman, 1930; Mayer, 1952) and the **scale count index**. The scale index is defined as the ratio of the free proximo-distal length of a scale to the diameter of the hair shaft. The scale count index is defined as the number of scales per unit (1 mm) length of hair shaft. Manuals on species characterization based entirely on the cuticle have been published (Adorjan and Kolenosky, 1969).

3.6.2   Classification of Cuticle

The classification criteria for cuticles have been described previously (Wildman, 1954; Brunner and Coman, 1974; Teerink, 1991). The classification of cuticles is based on the following four features:

- Scale position in relation to the longitudinal axis of the hair
- Shape of the scale margin
- Distance between the external margins of the scales
- Type of scale pattern

Prior to the detailed classification of scale and scale patterns, the scales can be broadly classified into two main types or groups: **coronal** (figure 3.5) and **imbricate** (see figure 3.17 through figure 3.20 and figure 3.22 through figure 3.28 later). A coronal scale will go around the hair shaft, completely encircling it. Coronal scales are found mostly in wool fibers or the under-hair. An imbricate scale does not go around the hair shaft and does not encircle it completely. Hence, several such scales are required to go around the circumference of the hair shaft at a particular region, to fully encircle it.

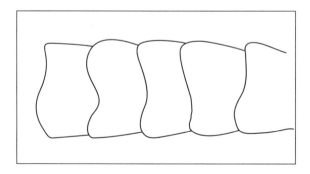

Figure 3.5 Coronal scales.

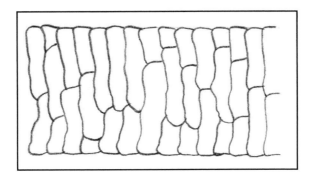

Figure 3.6 Transversal position.

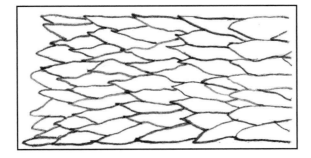

Figure 3.7 Longitudinal position.

Scale position and orientation in relation to the longitudinal axis of the hair: The scales can be described as transversal, longitudinal, or intermediate on the basis of their relation to the longitudinal axis of the hair.

1. **Transversal**: Scales are arranged at right angles to the longitudinal axis of the hair and their width is greater than their length (figure 3.6).
2. **Longitudinal**: The scales are arranged parallel to the longitudinal axis of the hair and their length is greater than their width (figure 3.7).
3. **Intermediate**: The width of the scales is approximately equal to the length of the scale (figure 3.8).

Shape of the scale margin: Scale margin represents the free distal end of the scale. There are five types of scale margins:

1. **Smooth margins**: The free distal end of the scale is without any indentations and appears as a smooth line (figure 3.9).

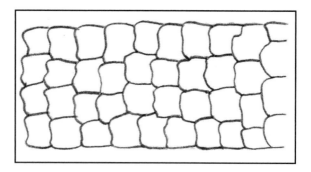

Figure 3.8 Intermediate position.

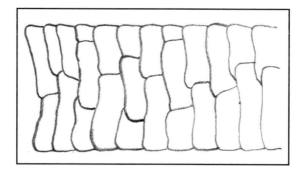

Figure 3.9 Smooth margins.

2. **Crenate margins**: The scale margins have shallow and relatively pointed indentations (figure 3.10).
3. **Rippled margins**: The indentations are deeper but have a relatively rounded profile as compared to crenate margins (figure 3.11).
4. **Scalloped margins**: The scale margin has a series of curves with rounded peaks and pointed troughs (figure 3.12).
5. **Dentate margins**: The distal free end of the scale has projections like teeth (figure 3.13).

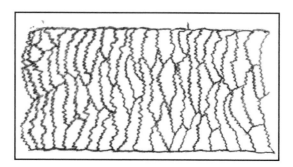

Figure 3.10 Crenate margins.

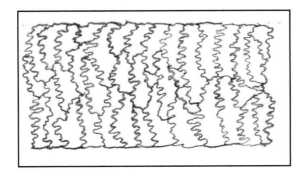

Figure 3.11 Rippled margins.

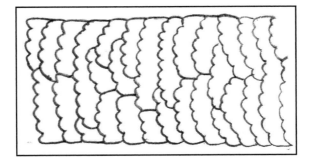

Figure 3.12 Scalloped margins.

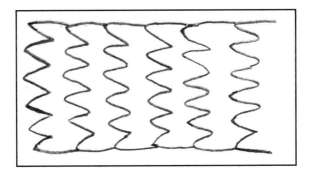

Figure 3.13 Dentate margins.

Distance between the external margins of the scales: The distance between scale margins describes the distance between the free ends (distal) of two adjacent scales along the longitudinal axis of the hair. The distance varies considerably amongst various species. The distance as such is not expressed in units but arbitrarily designated as **close**, **near**, and **distant** (figure 3.14 through figure 3.16). The width of the scales is apparently very

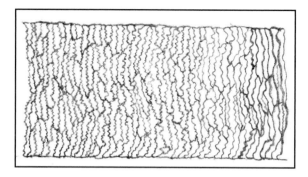

Figure 3.14 Close scales.

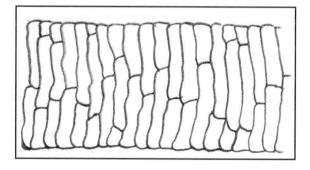

Figure 3.15 Near scales.

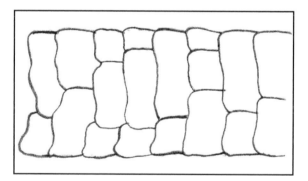

Figure 3.16 Distant scales.

great in comparison to their length when described as a **close** type, where the width of the **distant** scales is slightly greater than or equal to their length.

   Scale or cuticular patterns: The scale pattern is the arrangement of the scales along the longitudinal axis of the hair shaft. The scale pattern is particularly useful in the examination and comparison of imbricate hair types. These patterns vary along the length of the hair shaft from the proximal to the distal region along with the changes in the shapes of scales. The scale patterns show characteristic similarities and marked dissimilarities between related and unrelated species. The basic classification for cuticular patterns was provided by Wildman (1954), and more features were added by later workers (Brunner and Coman, 1974; Teerink 1991). The classification provided in this chapter is based on these studies.

The cuticular patterns can be classified into the following broad groups:

Petal pattern: The gross appearance of this pattern was described as being similar to the patterns formed by a series of overlapping petals of a flower (Wildman, 1954). In this group there can be four distinct sub-groups or patterns:

1. Broad petal: In this case the pattern comprises wide scales (figure 3.17).
2. Elongate petal: The scales are elongated (figure 3.18).
3. Diamond petal: In this pattern the scales give a diamond-like appearance due to the characteristic overlap (figure 3.19).

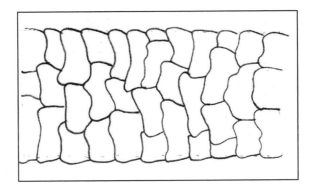

Figure 3.17 Broad petal pattern.

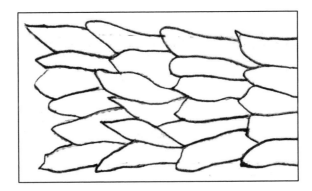

Figure 3.18 Elongated petal pattern.

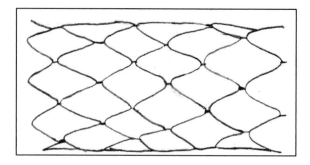

Figure 3.19 Broad diamond petal pattern.

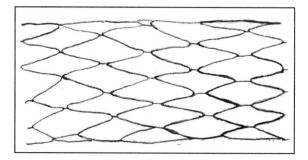

Figure 3.20 Narrow diamond petal pattern.

4. Narrow diamond petal: The scales have essentially a diamond petal configuration but are relatively long and narrow (figure 3.20). A narrow diamond pattern is commonly observed in the proximal region of hairs of Mustelids. Figure 3.21 shows a narrow diamond pattern as observed in the smooth coated Indian otter (*Lutrogale perspillata*).

Mosaic pattern: The name of the pattern is self-explanatory and applies to the gross appearance of the pattern (figure 3.22). In mosaic patterns the adjacent scales have rather straight margins (Teerink, 1991). Mosaic patterns can be further divided into two types:

1. Regular mosaic: A mosaic pattern with scales having visibly similar length and breadth (figure 3.22). This pattern is observed most clearly in Shahtoosh guard hair (figure 3.21).
2. Irregular mosaic: This pattern appears when the individual scales forming the mosaic pattern vary in their size and shape (figure 3.23).

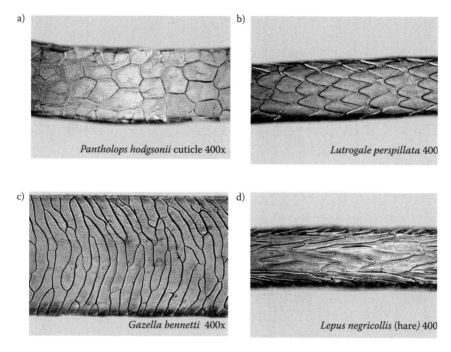

a) *Pantholops hodgsonii* cuticle 400x

b) *Lutrogale perspillata* 400

c) *Gazella bennetti* 400x

d) *Lepus negricollis* (hare) 400

Figure 3.21 Typical cuticle patterns observed in mammals. a) mosaic pattern, b) narrow diamond petal, c) regular wave pattern, and d) double chevron.

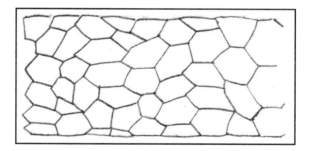

Figure 3.22 Regular mosaic pattern.

Wave pattern: In these patterns the distal edges or margins of the scales are undulating, appearing like a series of waves. Five different sub-groups or types can be identified in the wave pattern:

1. Regular wave: The name is self-explanatory and applies to the distal edge or margin of the scale showing regular and shallow troughs (figure 3.24). It is commonly seen among artiodactyl species (even toed ungulates such as the camel, goat, rhino, and horse). A regular wave pattern in the case of *Gazella* sp. is shown in figure 3.21.

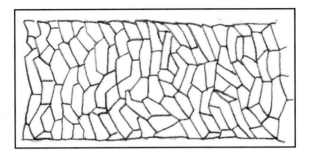

Figure 3.23 Flattened irregular mosaic pattern.

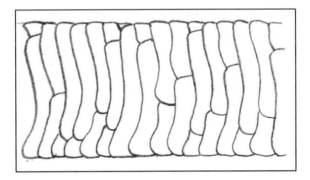

Figure 3.24 Regular wave pattern.

2. Irregular wave: In case of irregular waves, the troughs are irregular and deeper and as such the amplitudes of successive waves change considerably (figure 3.25).

3. Single chevron: In single chevrons either the crests or the troughs of the wave are pointed (Brunner and Coman, 1974). The pointed crests or troughs give a shape similar to English alphabet "V" (figure 3.26).

4. Double chevron: In double chevrons both the crests and the troughs are pointed and the formation of the "V" shape is in both directions (figure 3.27). Single and double chevron patterns are common in lagamorphs (rabbits and hares; figure 3.21).

5. Streaked wave: In certain wave patterns there is a longitudinal streak along the length of the pattern (figure 3.28). This occurs in hairs having longitudinal furrows on their shields.

Transitional pattern: The scale pattern changes along the length of the hair shaft from the proximal to the distal end, and there are points or regions where one type of scale pattern changes into another type. The patterns observed at such points or regions are called transitional patterns (figure 3.29).

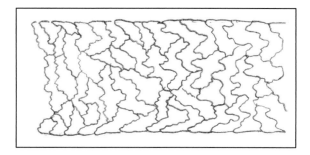

Figure 3.25 Irregular wave pattern.

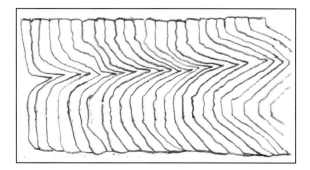

Figure 3.26 Single chevron pattern.

Figure 3.27 Double chevron pattern.

3.7   Cortex

The layer between the outer cuticle and the inner medulla is called the cortex. The cortex comprises dead cornified cells that are packed into a rigid and homogenous hyaline mass (Hausman, 1932). Cellular detail cannot be identified in the cortex due to the high cornification and close packing, and as such it does not have much value in species identifications using light microscopes.

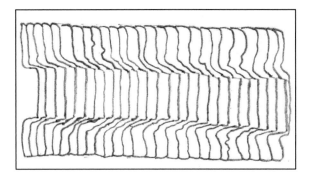

Figure 3.28 Streaked pattern.

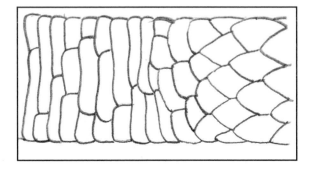

Figure 3.29 Transitional pattern.

The pigments that are responsible for the color of the hair are distributed in the cortex and may take the form of discrete granules, large amorphous masses, or diffuse stains (Hausman, 1930). The pigment granules in the cortex, in sharp contrast to the main cortical components, are of definite use in identification (Kirk and Thornton, 1974). The ratio of the cortex width to the hair diameter or thickness is of diagnostic use in species identification (measured in a cross-section slide).

3.8   Medulla

The medulla, or the core of the hair, varies widely along the species and is very valuable in species characterization and identification (Kirk and Thornton, 1974). It comprises shrunken dead cells with the spaces between the cells filled with air, and, as such, the medulla appears as a dark area under a light microscope in a whole mount (Brunner and Coman, 1974). This dark appearance obscures the view of the internal structure of the medulla and hence, in order to make the medulla transparent, it is necessary to remove the air by infiltration with some suitable medium like an organic solvent such as xylene.

The morphology and size of medulla vary significantly across species. The structure of the medulla is of great value in identifying the orders of the mammals from unknown hair samples, but it can rarely be much more specific. It is worthwhile to observe medulla structure with and without infiltration with a medium. In certain species the medulla may be absent.

3.8.1 Classification of Medullae

The medullae have been classified into four basic groups: **unbroken**, **broken**, **ladder**, and **miscellaneous** on the basis of the general shape and arrangement of cells and air spaces (Wildman, 1954; Brunner and Coman, 1974). These four major groups can be further divided into more descriptive categories. Margins of medullae have also been found useful in species identification and therefore medullae have been classified on the basis of their margin shape.

Unbroken medulla: In such cases the medulla or the core is continuous and unbroken along the length of the hair shaft. The diameter of the medulla may vary from narrow to very wide (Brunner and Coman, 1974). This can be further classified as lattice, aeriform lattice, and simple types:

1. Lattice medulla: The shrunken cells of the medulla form a lattice or a network, enclosing air spaces of various shapes (Brunner and Coman, 1974). Lattice medullae are the most frequently observed unbroken medullae. Lattice medullae may further be designated as **narrow** or **wide** lattice medullae (figure 3.30 and figure 3.31), depending on the **medullary index** (ratio of medullary thickness to hair thickness or diameter). Medullary index values below 0.5 represent narrow lattice medulla. Wide medulla lattices are commonly seen amongst Cervids (deer species). Figure 3.32 shows a wide lattice medulla as seen in *Axis porcinus* (hog deer) and a narrow lattice medulla in the case of *Capricornis sumatraensis* (Indochinese serow).

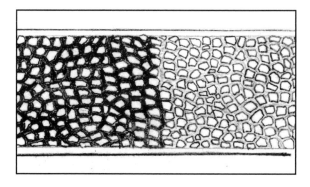

Figure 3.30 Wide lattice medulla.

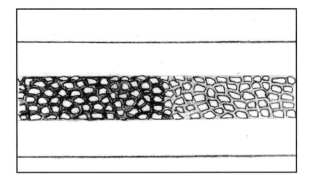

Figure 3.31 Narrow lattice medulla.

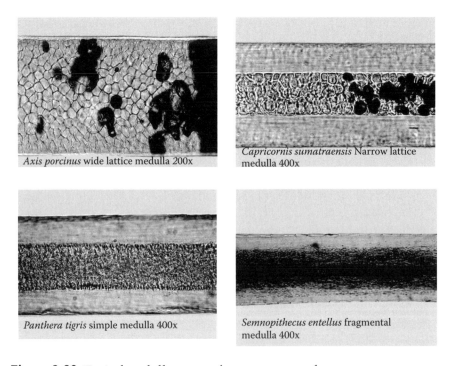

Axis porcinus wide lattice medulla 200x

Capricornis sumatraensis Narrow lattice medulla 400x

Panthera tigris simple medulla 400x

Semnopithecus entellus fragmental medulla 400x

Figure 3.32 Typical medulla pattern for some mammals.

2. Aeriform lattice: In this type of medulla the air spaces form a lattice or a network, enclosing the shriveled medullary cells (Brunner & Coman, 1974). Aeriform lattice medullae are commonly observed in rodent hairs. This type of medulla can be further designated as **narrow** or **wide** lattice medullae (figure 3.33 and figure 3.34), depending on the **medullary index**. Medullary index values below 0.5 represent narrow aeriform lattice medullae.

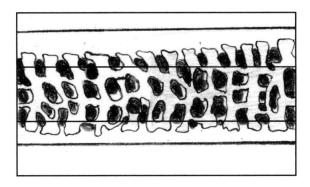

Figure 3.33 Wide aeriform lattice medulla.

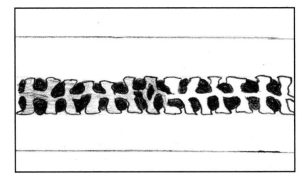

Figure 3.34 Narrow aeriform lattice medulla.

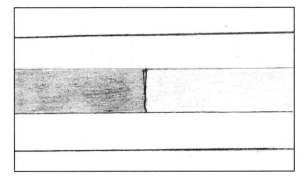

Figure 3.35 Simple medulla.

3. Simple medulla: This type of medulla has no visible structure or pattern and gives an amorphous appearance (figure 3.35). In contrast to other types of patterns, little can be revealed by infiltration with xylene or oil. A simple medulla may be narrow or wide, but most are of a medium width. These are frequently seen in larger cats (such as *Panthera tigris*) (figure 3.32).

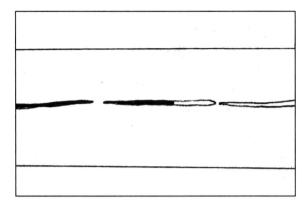

Figure 3.36 Interrupted medulla.

Broken or discontinuous medulla: In some hairs the medullae are not continuous along the length of the hair, as there are interruptions by small or long sections of cortex. Such medullae are common among primate species. Broken medullae can be further divided into two sub-groups: **interrupted** and **fragmented**.

1. Interrupted: The breaks in the medulla column are due to short sections of cortical material (figure 3.36 and figure 3.32).
2. Fragmental: The breaks in the medulla column are due to long sections of cortical material (figure 3.37 and figure 3.38).

Ladder medulla: In this type of medulla there are one or more rows of air spaces that give an appearance of a ladder. The air spaces appear black in color without infiltration with xylene or oil and are very distinctly visible. After infiltration they appear light in color and become less distinct. These can be further divided into two distinct categories:

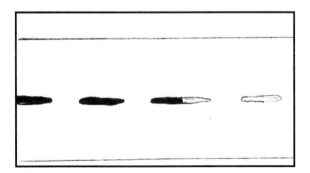

Figure 3.37 Fragmental medulla.

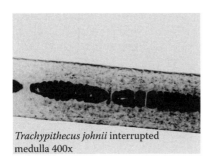

Trachypithecus johnii interrupted medulla 400x

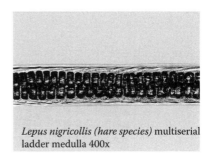

Lepus nigricollis (hare species) multiserial ladder medulla 400x

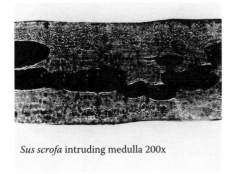

Sus scrofa intruding medulla 200x

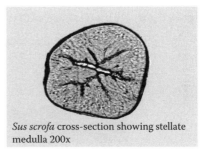

Sus scrofa cross-section showing stellate medulla 200x

Figure 3.38 Medulla characteristics of selected species.

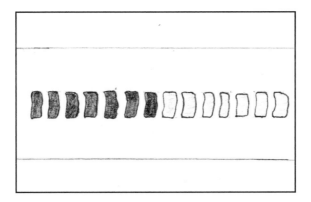

Figure 3.39 Uniserial ladder medulla.

1. Uniserial ladder: This type of medulla has a single row of air spaces (figure 3.39). This pattern is commonly observed in cases of under-hairs of most of the felid and mustelid species.
2. Multiserial ladder: There are two or more rows of air spaces in this type of medulla (figure 3.40). This type of medulla is peculiar to the order Lagomorpha (rabbits and hares; figure 3.38).

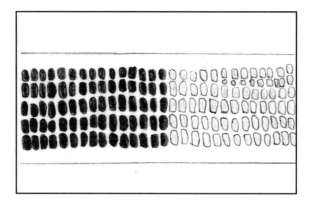

Figure 3.40 Multiserial ladder medulla.

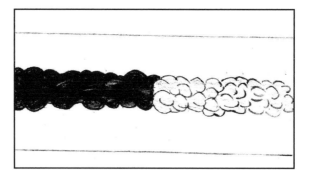

Figure 3.41 Globular medulla.

Miscellaneous type: This is a collection of the more uncommon types of medullae (Brunner & Coman 1974). Three types have been recognized:

1. Globular: This type of medulla comprises an aggregation of air spaces that are (as the name suggests) globular in shape (Brunner & Coman 1974) (figure 3.41).
2. Stellate medulla: This type of medulla is a simple type but has sharp projections radiating into the cortex (figure 3.42), and hence in a cross-section the medulla appears star-like (stellate) (figure 3.38).
3. Intruding medulla: These medullae show irregular projections into the cortex. These projections may be in any direction and, as such, the medullae are very unevenly distributed around the center of the hair (figure 3.43 and figure 3.38).

Stellate and intruding medullae are frequently seen in hair of pigs (family Suidae).

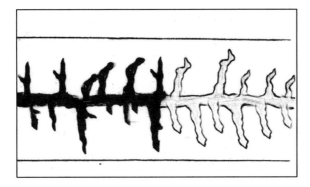

Figure 3.42 Stellate medulla.

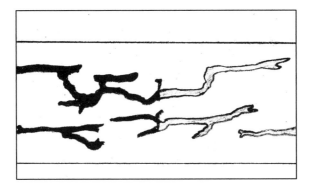

Figure 3.43 Intruding medulla.

Classification on the basis of margin of medulla: In addition to the previously mentioned features, the margins of medullae in the widest portion of the hair (shield) have been used for species characterization (Teerink, 1991). These features can be best observed before the infiltration of medullary air spaces with xylene or oil. On the basis of margins three categories are recognized:

1. Straight margins: The margins of the medulla are straight without any cortical intrusions (figure 3.44).
2. Fringed margins: Small protrusions from the medulla extend into the cortex (figure 3.45).
3. Scalloped margins: The outline of the medulla is represented by a series of rounded or convex projections (figure 3.46).

Cross-section: As discussed, in the hair profile, mammalian hairs show significant variations in their shapes. This variation of the hair structure morphology is revealed more in their cross-sections. For hairs with cylindrical shapes, the cross-section will be essentially circular in outline, and

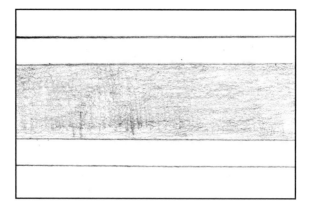

Figure 3.44 Straight medulla margins.

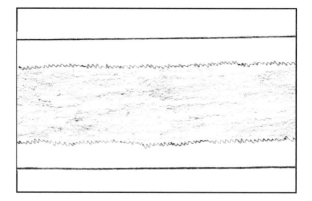

Figure 3.45 Fringed medulla margins.

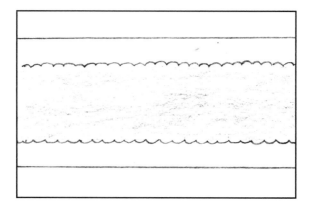

Figure 3.46 Scalloped medulla margins.

for complex shapes the cross-sections are of very distinct shapes. The shapes of the cross-sections and their dimensions can be of value in species identification. Taking a cross-section allows for the calculation of the ratio of the medulla and the cortex, with respect to hair thickness. Cross-sections from the widest portion of the shield are most informative for species identification (Teerink, 1991); however, cross-sections from other regions of hair should also be studied for a complete characterization or identification of the species.

The cross-sections of hairs can be of various shapes, and the outlines of some of the important shapes encountered in mammals are given in figure 3.47.

3.9 Hair Pigments

Pigment, although not of much value in species identification, sometimes may provide corroborative information for this purpose. The pigment can be studied on the basis of the pigment distribution, pigment color, and pigment type.

3.9.1 Pigment Distribution

The pigment may be present in the cortex or medulla or in both, but rarely in the cuticle (Benedict, 1957). The arrangement of the pigment granules, which may be peripheral or central, is used in species identification and can be best seen in a cross-section of the hair. In certain species the pigment distribution along the length of hair is very informative. Such variable distributions of pigments give a banded appearance to the hair (comprising alternate light and dark bands) (figure 3.48). These banding aspects have been used in species identification keys (Stains, 1958; Mayer, 1952). Such banding patterns are common in mongoose species (*Herpestids*), civets (*Viverrids*), and some rodents.

3.9.2 Pigment Color

The color of hair is due to melanin (Eumelain and Pheomelanin) pigments. Hairs display a wide variety of color hues that can be used in species identification for species with a particular hair coloration.

3.9.3 Pigment Type

The pigment in the hair may take the form of granules, large amorphous masses (aggregate), or a diffuse stain. These terms are self-explanatory and have some value in species identification. Pigment granules or aggregate pigments may be arranged in well-defined areas or arranged in streaked patterns

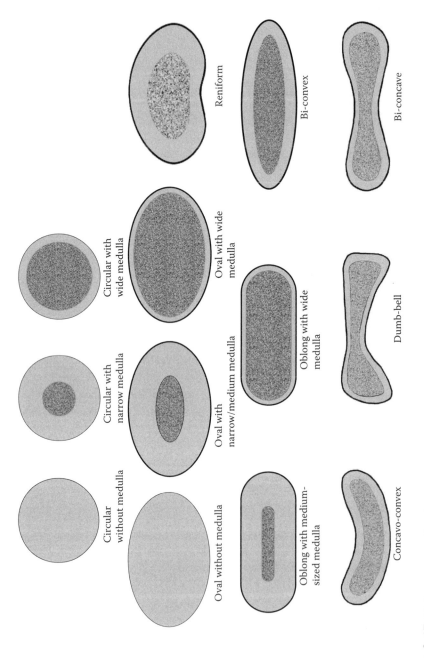

Figure 3.47 Hair cross-section types.

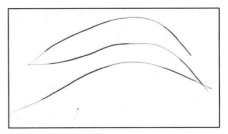

Mongoose *(Herpestes sp.)* hair showing banding pattern due to differences in pigment distribution

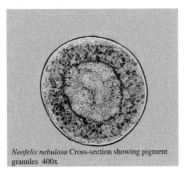

Neofelis nebulosa Cross-section showing pigment granules 400x

Figure 3.48 Hair pigments.

(Brunner and Coman, 1974). An example can be found in the cross-section of the clouded leopard (*Neofelis nebulosa*), which is shown in figure 3.48.

As the pigment distribution and color vary considerably along the length of the hair shaft, along body parts of the same individual, and across individuals of the same species, and as the pigment is subject to color mutations and fading, it is not a very reliable tool for species characterization and is of limited use.

3.10 Techniques for Studying Hair Structure

3.10.1 Cleaning of Hair Samples

For the study of hair structures to characterize or identify species, it is necessary that the hair samples are clean and dirt-free. Hair samples obtained as reference collection material from museums, etc., and hair evidence obtained for species identification normally have a great deal of unwanted dirt sticking to their surfaces, which hinders the preparation of good-quality slides. In order to remove dirt from the surface of the hair, it is necessary to wash them with distilled water and a mild detergent to remove inorganic debris. Hairs may be rinsed in distilled water for 4–5 min, depending on the amount of dirt. Once the hairs have been rinsed in water, organic solvents like carbon tetrachloride or isopropyl alcohol are used to remove the organic debris such as grease and wax from the surface of the hairs.

3.10.2 Hair Profile

The profile of the hair can be observed easily by preparing a whole mount of the hair on a microscopic slide using a mounting media and a cover slip. If a permanent slide is required for the reference material, Di-n-butyl Phthalate in Xylene (DPX) can be used as mounting media. For hair samples obtained as case exhibits, temporary mounts are prepared such that the limited evidence can be used in further examinations of the cuticle and medulla and for making cross-sections if required. A stereomicroscope is the best way to view a whole hair.

3.10.3 Preparation of Cuticular Slides

The surface of the hair can be studied using light microscopy or an SEM. The former method has the advantage of its simplicity and cost effectiveness. The latter method (SEM) provides very high resolution images of the hair surface but is less cost effective and is subject to the availability of such a facility. Moreover, the resolution and magnification provided by the light microscope can provide adequate information needed for species identification from hair. Both methods are discussed separately.

Light microscopy method: As the compound or comparison microscopes used for hair examination are essentially transmitted light microscopes, it is not possible to view the cuticle on a whole mount as it is almost transparent and is a very thin layer of cells. Only the cortex and medulla are visible in a whole mount using light microscopy. Hence, to view the cuticular structure of hair, a **cast** or **impression** of hair is made and viewed under the microscope. Most hair studies have been carried out using these techniques. There is a clear distinction laid down between "cast" and "impression" (Wildman, 1954). In hair "casts" essentially only a part of the hair circumference is represented, whereas with "impressions" the complete circumference of the hair at a particular region is represented. The hair impression is obtained by rolling the hair over a surface of a suitable medium on a slide such that the entire circumference of the hair can be viewed (Brunner and Coman, 1974). This method is not commonly practiced due to its time-consuming nature and is not applicable to hairs with highly flattened shields. Further, the hair may be damaged by such rolling.

In the case of the hair "scale cast" method, a hair is placed or pressed on the surface of a suitable material such that the surface structure of the hair gets reproduced as a three-dimensional cast. This cast can be viewed under light microscopy to observe the cuticular structure of the hair. There are several media, like gelatin, polyvinyl acetate (PVA), celluloid, and nail polish, that can be used for preparing scale casts. The best media are gelatin and polyvinyl acetate, because they reproduce the cuticular structure of hair very effectively. A solution of gelatin (10–20%) is prepared in distilled water. Alternatively, a 50% solution of PVA in distilled water can be prepared for the collection of scale casts (Brunner and Coman, 1974; Teerink, 1991). These media can be prepared in a 100-ml borosilicate glass beaker. Gelatin and PVA require heating to dissolve in water, which can be done with a hot plate or a microwave heater. Once the gelatin or PVA has dissolved, a small amount of methylene blue dye can be added to the media as the blue color enhances the view of the scale casts under the microscope. Nail polish can be used when such media are not available. The efficacy of reproduction of the scale pattern on the surface of the casting media depends on the uniformity of the thickness of the media; hence, emphasis must be laid on obtaining a fine and uniform film of the casting media. A fine and uniform film of the casting media is made on a clean microscopic glass slide (50 × 20 × 0.2 mm) with the help of a glass rod or a fine flat brush in a single stroke along the length of the slide surface. Any excess mounting media on the slide can be drained from the slide by tilting the slide and wiping the edges with tissue or filter paper. The slide is placed on a horizontal surface and the hair samples are placed one by one on the slide with the help of tweezers. The root ends of some hairs and the tip ends of the others should be left hanging free such that it is easy to remove the hair from the glass slide once the medium has dried (figure 3.49). The medium dries in 20–30 min at room temperature and the hairs are plucked off with the help of tweezers. The cast left behind by the hair is viewed under a high-magnification (100× to 400×) microscope to view the cuticular pattern.

As discussed earlier, hairs rarely have a perfect cylindrical outline and show flattening at certain regions (shield). Further, the hair might be kinky or wavy in outline. In such cases it becomes extremely difficult to place the hair properly on the surface of the medium. In such cases care should be taken that the widest portion of the hair is placed flat on the medium surface to obtain a proper cast. Long and kinky hair may be cut into small pieces to obtain casts, representing the different regions of long and kinky hair.

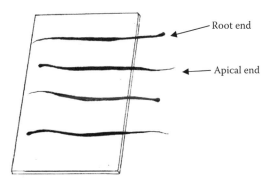

Figure 3.49 Placing of hair on a slide to obtain cuticular scale casts.

Scale casts made in gelatin and PVA cannot be stored for a long period of time and hence photographic documentation of the cuticular structure is required.

Scanning electron microscopy (SEM) method: Instead of light, the samples are viewed with a beam of electrons that scan the surface of interest in an array of picture points. The image obtained is essentially black and white (figures 3.3b and c), although this depends on the type of monitor and detection system used. The resolution and magnification are extremely high in comparison to a conventional light microscope. SEM has been used by several investigators for cuticular studies (Vogel and Kopchen, 1978; Dziurdzik, 1978; Short, 1978; Hutterer and Hurter, 1981; Rollins and Hall, 1999; Phan et al., 2000).

For examining the hair samples under SEMs, the hair samples need to be cut into small pieces, about 0.5 to 1.0 cm in length. These samples are mounted onto an aluminum mounting stub with the help of double-sided adhesive tape. A piece of tape (1 cm^2) is pasted flat on the stub surface and the hair samples are placed with the help of tweezers on the upper surface of the tape and gently pressed in place. The hair samples as such cannot be examined under SEM as their surface needs to be coated with an element capable of being detected by the electron beam. This is achieved by coating a thin film of conducting material (gold or palladium) on the surface of the hair samples under low pressure (10^{-6} torr). The specimen is held on an anode base, a few centimeters beneath a cathode to which is fixed the material to be used in the coating process (gold or palladium). The vacuum vessel, in which the samples are housed, is flushed with an inert gas (usually argon) and then pumped to a vacuum of about 5 pa. The application of a few kilovolts between the two electrodes begins the coating process.

Collision of the inert gas atoms with energetic free electrons, which are being accelerated toward the anode, causes ionization. The resulting positive ions are then themselves accelerated toward the target cathode, which they strike with sufficient force to dislodge atoms by momentum. These atoms are then deposited on all the surfaces within the vacuum vessel, including the specimen. Once the samples are coated with gold or palladium they can be viewed under an SEM.

3.10.4   Preparation of Medulla Slides

Medulla slide preparation is perhaps the easiest to prepare, as the procedure is very similar to the preparation of a whole mount. As discussed earlier, the medulla is filled with air and thus appears as a black region in the core of hair. It is necessary to remove the air to get a detailed view of the medulla. This is done by infiltrating the medulla with xylene or paraffin oil, which removes the air and makes it transparent when viewed through the microscope. To allow this to happen, the hair samples are cut at various positions such that xylene or paraffin oil can percolate into the medulla.

Different investigators have suggested methods for medulla slide preparation. A good method is to place the hair on a microscopic glass slide and fix it in place with drops of polyvinyl chloride acetate glue, as described by Teerink (1991). The hair is cut along the entire length with a razor blade. These fixed hair segments are mounted in paraffin oil that slowly percolates into the medulla and makes it transparent and ready for examination under the microscope.

The simplest method is to cut the hair sample into small pieces of about 0.5 cm to 1.0 cm in length with a razor blade and immerse them in xylene kept in a Petri dish. The medulla becomes fully infiltrated with xylene in 1–3 h and the hair can be mounted in DPX mounting media and examined under the microscope. This is a very reliable method for obtaining high-quality medulla slides. The shape and size of medulla should be studied along the length of the hair, but the shield region is most informative. Figure 3.50 shows the medulla structure of a hair from a deer sp. (*Cervus unicolor*). Note that without treatment with xylene, the medulla appears as a dark region (figure 3.50a), and after treatment with xylene, the medulla structure is clearly visible in the same hair (figure 3.50b).

This method works perfectly for nonpigmented or lightly pigmented hair. However, in the case of certain species (some primates, black bears, etc.) where the hairs are heavily pigmented, i.e., the cortex has a large amount of dark pigments, it is not possible to view the medulla even after treatment with xylene or oil due to the opaqueness of the cortex. In order to view the medulla, the cortex must be made transparent by chemical treatment. This is done by immersing dry hairs in a Petri dish containing hydrogen peroxide, to which a few drops of ammonia solution are added. The Petri dish is kept covered and the hair samples are kept immersed until the desired level of lightness is obtained. These hairs can then be dried and treated with xylene or oil to prepare medulla slides. Figure 3.51 shows the medulla structure of

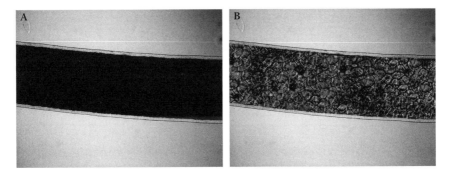

Figure 3.50 (a) Medulla visible as a dark region in a hair of *Cervus unicolor* (without treatment with xylene). (b) Medulla structure clearly visible after treatment with xylene in the same hair.

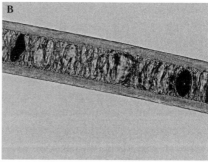

Figure 3.51 (a) A heavily pigmented hair of *Bos gaurus* revealing nothing (without treatment with hydrogen peroxide and ammonia). (b) Cortex and medulla structures are clearly visible after treatment with hydrogen peroxide and ammonia followed by percolation of the medulla with xylene in the same hair.

a hair from a bovid species (*Bos gaurus*). Note that without treatment with hydrogen peroxide and ammonia, the whole hair appears dark, revealing almost nothing (figure 3.51a), and after treatment with hydrogen peroxide and ammonia, followed by percolation of the medulla with xylene, the cortex and medulla are clearly visible in the same hair (figure 3.51b).

3.10.5   Preparation of Cross-Sections

Preparation of good-quality cross-sections of hairs is necessary for species identification as cross-sections provide valuable information about the genus and species. Cross-sections should be obtained from various parts along the entire length of the hair shaft. The cross-section from the shield region is, however, considered to be the most informative. Several methods are available for the preparation of hair cross-sections. The methods are mostly based on cross-sectioning manually or with a microtome.

A variety of microtomes is available and previous publications on the use of hair cross-section have been published (Wildman, 1954; Stoves, 1951). To obtain a cross-section, a sample is embedded in a resin or wax and sections are cut with steel blades of the microtome. Very recent advanced microtomes use a laser for cutting sections of histological material, and the same can be applied for hair samples. However, despite the various mechanical methods available for cross-sectioning, the manual method of section cutting has the advantage of its simplicity and effectiveness. One important factor that has a bearing on the manual method is the skill of the examiner; hence, it requires a great deal of practice to refine the necessary skills.

Various investigators have described a variety of manual methods for cross-sectioning. Mathiak (1938) used a celluloid solution and balsa wood for cross-sectioning. A plate method described by Ford and Siemens (1959) was used by various workers (Brunner and Coman 1974). A cellulose acetate layer–based method was described by Teerink (1991).

We describe a simple alternative method that can be used to obtain very good-quality cross-sections with minimal investment. This method requires a straw pipe, mounting wax, and a razor blade for preparing the cross-sections. A few hairs are inserted into a straw pipe to keep them as straight as possible (figure 3.52). While maintaining the position of the hair, molten wax is sucked carefully into the straw pipe from the opposite end. Once the wax rises past the hair samples, the straw pipe is constricted with fingers in order to stop the wax from flowing out. The straw pipe is then inverted and the wax is allowed to solidify. Brief refrigeration may assist in the solidification if the environment is too warm. Once the wax is solidified completely, the straw pipe is cut open with a blade and the stub of hair fixed in wax can be used to cut cross-sections of the hair with a razor blade. The cut sections can be placed on a microscopic slide, de-waxed using xylene, and viewed under a microscope after placing a cover slip. This requires good skill, which can be learned in 2–3 days.

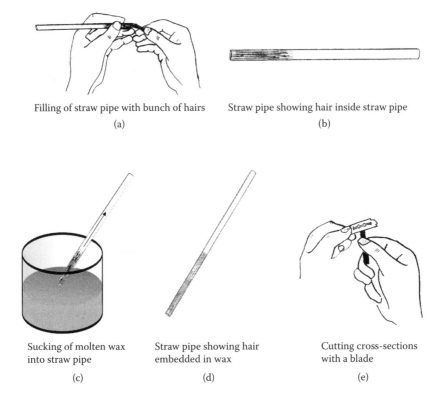

Filling of straw pipe with bunch of hairs
(a)

Straw pipe showing hair inside straw pipe
(b)

Sucking of molten wax into straw pipe
(c)

Straw pipe showing hair embedded in wax
(d)

Cutting cross-sections with a blade
(e)

Figure 3.52 (a-e) Preparation of a cross-section.

3.11   Features to be Examined during Hair Analysis

Case exhibits in wildlife forensic investigations may involve a few hairs, a complete hide, part of a hide, or finished products like fur coats and caps or shawls. Each case scenario is bound to be different, and hence there will be subtle differences in the strategy to be followed when a microscopic examination is to be carried out for species identification. A simple tabular format is provided in Appendix 1 at the end of this chapter that explains clearly all features that are to be examined and how the data have to be recorded and presented. Since a visual comparison of microscopic hair characteristics of a reference hair sample and case sample helps to provide a better impression about the analysis, it is always worthwhile to have a tabular representation, depicting microphotographs of reference species hair and the case exhibit hair (Appendix 2).

3.12   Conclusions

We have tried to provide a very generalized and fundamental approach for species characterization and identification from hair using microscopy. As the method is very subjective, training and experience are essential. A comprehensive repository of hair samples of various species of interest is also essential. This repository may reflect the range of samples that are encountered by the laboratory. Hence, it is important to have trichotaxonomists specializing in species of a particular region, and there should be a good exchange of knowledge and a database on microscopic hair characteristics among these experts.

The cost effectiveness and the non-destructive nature (except for the cross-sectioning) make hair characterization with microscopy a very useful tool for undertaking species identifications. Microscopic examination by an experienced trichotaxonomist can help in narrowing down the search to a few taxonomic groups of animals, even if the hair evidence is very minimal. Examination of exhibits under microscopes will always provide good information and investigation leads in a very non-destructive manner. Hence, it is always worthwhile to examine evidence material under a suitable microscope.

References

Adorjan, A. S. and Kolenosky, G. B. (1969). A manual for the identification of hairs of selected Ontario mammals. *Res. Report (Wildlife)*. Dept. Lands and Forests. Ontario.

Appleyard, H. M. (1978). *Guide to Identification of Animal Fibres*. 2nd edn., Wool Industries Research Association, Leeds.

Benedict, F.A. (1957). Hair structure as a generic character in bats. *Univ. Calif. Pub. Zool.* 59: 285–548.

Brunner, H. and Coman, B. J. (1974). *The Identification of Mammalian Hair.* Inkata Press, Melbourne.

Dreyer, J. H. (1966). A study of hair morphology in the family Bovidae. *Onderstepoort. J. Vet. Res.* 33: 379–472.

Dziurdzik, B. (1978). Histological structure of hair in the *Gliridae (Rodentia). Acta Zool. Cracov.* 23: 1–10.

Ford, J. E. and Simmens, S. C. (1959). Fibre section cutting by plate method. *J. Text. Inst. Proc.* 50: 148–158.

Hardy, J. I. and Plitt, T. M. (1940). An improved method for revealing the surface structure of fur fibers. *U.S. Dept. Interior Wildlife Circ.* 7, 10p.

Hausman, L. A. (1920). Structural characteristics of the hair of mammals. *Am. Nat.* 54: 496-523.

Hausman, L. A. (1924). Further studies of the relationships of the structural characters of mammalian hair. *Am. Nat.* 58: 544-557.

Hausman, L. A. (1930). Recent studies of hair structure relationships. *Scient. Monthly.* 30: 258–277.

Hausman, L. A. (1932). Cortical fusi of the mammalian hair shafts. *Am. Nat.* 66: 461–470.

Hausman, L. A. (1944). Applied microscopy of hair. *Scient. Monthly.* 59: 195–202.

Hutterer, R. and Hurter, T. (1981). Adaptive Haarstrukturen bei Wasserspitzmausen (*insectivora, Soricinae). Z. Saugetierk.* 46: 1–11.

Kirk, P. L. and Thornton. J. (Ed.) (1974). *Criminal Investigation.* 2nd edn., Wiley, New York.

Mathiak, H. A. (1938). A key to the hairs of mammals of Southern Michigan. *J. Wildl. Mgmt.* 2: 251–268.

Mayer, W. V. C. (1952). The hair of Californian mammals with keys to dorsal guard hairs of Californian mammals. *Amer. Midl. Nat.* 48: 480–512.

Moore, T. D., Spence, L. E. and Dugnolle, E. E. (1974). *Identification of the Dorsal Guard Hairs of Some Mammals of Wyoming.* Game and Fish Dept. Wyoming.

Phan, K. H., Wortmann, G. and Wortmann, F. J. (2000). Microscopic Characteristics of Shatoosh and Its Differentiation from Cashmere/Pashmina. *Proc. 10th Int. Wool Text. Res. Conf.*, Aachen, SF-2: 1–16.

Rollins, C. K. and Hall, D. M. (1999). Using light and scanning electron microscopic methods to differentiate Ibex goat and Tibetan antelope fibers. *Text. Res.* 69: 856–860.

Short, H. L. (1978). Analysis of cuticular scales on the hair using the scanning electron microscope. *J. Mammal.* 59: 261–267.

Stains, H. J. (1958). Field key to guard hair of Middle Western furbearers. *J. Wildl. Mgmt.* 22: 95–97.

Stoves, J. L. (1951). *Fibre Microscopy.* National Trade Press, London.

Teerink, B. J. (1991). *Atlas and Identification Key on Hair of West-European Mammals.* Cambridge University Press.

Vogel, P. and Kopchen, B. (1978). Besondere Haarstrukturen der Soricidae (*Mammalia, Insectivora*) und Ihre Taxonomische Deutang. *Zoomorphology* 89: 47–56.

Wildman, A. B. (1954). *The Microscopy of Animal Textile Fibres.* Wool Industries Research Association, Leeds.

Appendix 1   Features to be Assessed during Microscopic Examination of Hair

Wildlife Forensic Laboratory						
Case number :		Place:				
Case exhibit:		Type of examination:				
Date:		Page number:				
Features	**Hair number**					
A) Macroscopic		1	2	3	4	5
1. Color						
2. Profile						
3. Length						
4.Thickness at the widest portion or shield						
B) Microscopic						
1.1. **Cuticle** (*proximal region*)						
a) Scale position/orientation						
b) Scale margin						
c) Scale distance						
d) Scale pattern						
1.2. **Cuticle** (*medial region*)						
a) Scale position/orientation						
b) Scale margin						
c) Scale distance						
d) Scale pattern						
1.3. **Cuticle** (*distal region*)						
a) Scale position/orientation						
b) Scale margin						
c) Scale distance						
d) Scale pattern						
2. **Medulla**						
a) Type						
b) Medullary index						
c) Medulla margins						
3. **Cross-section**						
a) Shape						
b) Medulla configuration						
4. **Pigment**						
a) Type						
b) Distribution						
Notes:						
Examined by	Signature	Name		Designation		
Counter examined by	Signature	Name		Designation		

Appendix 2   Images of Microscopic Hair Characteristics for Reference and Case Hair

Wildlife Forensic Laboratory		
Case number : Place:		
Case exhibit: Type of examination:		
Date: Page number:		
Microscopic Feature	Image of Reference Hair (magnification x)	Image of case hair (magnification x)
1.1. **Cuticle** (*proximal region*)		
1.2. **Cuticle** (*medial region*)		
1.3. **Cuticle** (*distal region*)		
2. **Medulla**		
3. **Cross-section**		

Species Identification Using DNA Loci

4

ADRIAN LINACRE AND SHANAN S. TOBE

Contents

4.1 Introduction

The aim of this chapter is to describe the DNA tests that are used currently to identify an unknown sample as being of a particular species. This is a prime question in wildlife investigations, where the question "what species did this sample come from?" is frequently asked.

Species identification is a requirement for a large number of different reasons. Customs officers require the ability to identify species (or products derived from those species) being poached or traded contrary to national and international legislation. Animal interference after death can be confused with pre-mortem injuries, which can arouse suspicion of violence prior to death [1]. Rodents [2] and canids [3] frequently alter remains. Other needs include determining the species of origin of bloodstains [4]; identification of the components of meats (for religious purposes or to identify poached species that are out of season or endangered, or where a quota has been exceeded) [4]; and identification of human and animal components of commingled remains from mass disasters, fires, cremations, and domestic crimes [5].

It is important to be able to clearly identify animal species to determine population numbers, diversity, and distribution. Non-invasive sampling methods for species identification, such as identification from scat [6], are required. A sensitive, quick, reliable, and accurate means is needed to clearly identify animal species present in samples [4], including mixtures. Although new species are continually being discovered, the number of species disappearing is at a far greater rate. For this reason, conservationists are interested in preserving and protecting native and endangered species and stopping the introduction of invasive species. International legislation (stipulated by CITES) covering the illegal trade of endangered species stipulates which species are protected. Chapter 2 outlined the role of CITES in the enforcement of trade of protected species with a range of species listed in CITES Appendices I, II, and III. National legislation within many countries may stipulate the protection of listed species; for instance, this includes particular bird species in the United States and deer species in India. A prime question asked of a forensic science laboratory is therefore to determine which species is present and if this species is protected by either international or national legislation.

To be able to enforce such legislation, it is necessary to be able to define a species.

4.2 Definition of a Species

There are a number of definitions of a species, and currently the two main definitions are biological species and phylogenetic species [7, 8]. The biological species definition is based on gross morphological features, whereas the phylogenetic species definition represents the relationships between organisms as revealed by their evolutionary history [9].

An example of the definition of a species from the *Oxford English Dictionary* is "a group of living organisms consisting of similar individuals capable of exchanging genes or interbreeding." This corresponds with the biological definition of species. To expand, the biological definition of species assumes that each species is reproductively isolated (genes are passed within a species but not beyond) [7, 10]. One species cannot breed with a different species and produce viable young. The problem with any definition is that there are exceptions. This concept holds true when dealing with animal hybrids such as mules (which are infertile because horses and donkeys have different numbers of chromosomes, so the mule offspring will have an odd number) but runs into problems when dealing with hybrids of plants (which are able to reproduce) [7]. However, wolves, coyotes, and dogs can mate with each other and share the same number of chromosomes [11]. Even though mating is possible between these species, it is precluded by social factors and size differences, meaning that mating is a rare occurrence [11].

The phylogenetic definition of species relies on genetic markers (RNA, DNA, or protein) thought to be characteristic of a species [7, 9]. This is being widely used in taxonomy, biodiversity, and evolutionary studies [8]. Even with the advent of DNA and sequencing technologies, there is still no consensus as to how many genetic variations constitute a separate species.

4.2.1 Nomenclature

Normally all the members of the same species share a similar physical appearance, although this may be different between the sexes. It was this physical appearance that led Carl Linnaeus to develop a binomial Latin name for all known species. His system assigns a one-word name to a genus, to which more than one closely resembling species may belong, and a two-word name, with the first word being the same as for the genus, to a species. This system of species identification based on physical appearance was satisfactory for many years, but the advent of genetics has altered the classifications of some species. Further, there are sub-species (varieties within a species) where there may be phenotypic differences but individuals are capable of interbreeding and producing viable young. A well-known example of how genetics has affected the taxonomy of a species is the giant panda (*Ailuropoda melanoleuca*), which was originally considered as a bear but subsequently grouped with the red panda (*Ailurus fulgens*) although it shares much of its phenotype, and also its behavior and method of reproduction, with members of the bear family. The giant panda is currently placed within the bear family (*Ursidae*).

Tigers are used here as an example to demonstrate species and taxonomy. There are five currently living (extant) sub-species of tiger. The species name for tiger is *Panthera tigris* (note a species name is written in italics). The five sub-species are *P. tigris tigris* (Bengali), *P. tigris altica* (Siberian), *P. tigris corbetti* (Indo Chinese), *P. tigris amoyensis* (Amoy), and *P. tigris sumatrae* (Sumatran). All tigers are members of the Panthera genus. This genus includes lions (*Panthera leo* and *Panthera persica*), leopards (*Panthera panthera*), pumas (*Puma concolor*), and other large cats. All members of the Panthera genus belong to the Felidae family. This also includes the domestic cat (*Felis catus*). Going one category higher, all the Felidae belong to the order Carnivora, which also contains dogs (family Canidae) foxes (family Vulpes), and bears (family Ursus), among others. All the carnivores are members of the class Mammalia.

An example is given in table 4.1 of the classification of wolf, dog, fox, cat, and human. All are members of the class Mammalia and all are members of

Table 4.1   A Demonstration of the Increased Sequence Variation in the Cyt b Gene as Species Become More Distant Taxonomically

	Wolf	Dog	Fox	Cat	Human
Phylum	Chordata				
Class	Mammalia				
Order	Carnivora				Primate
Family	Canidae			Felidae	Hominid
Genus	*Canus*	*Canus*	*Vulpes*	*Felis*	*Homo*
Species	*lupus*	*lupus*	*vulpes*	*catus*	*sapiens*
Sub-species		*familiaris*			

Note: The first column lists the terms used in taxonomy starting at the bottom with the species name and ending with the phylum Chordata, into which all animals with a central nervous canal at the back (the spine in mammals) are grouped. There are two classifications higher, and those are kingdom and domain.

the order Carnivora, except humans, which are in the order Primate. Wolf, dog, and fox are members of the family Canidae, followed by genus Canus. Dog is here classified as a sub-species of wolf (*Canus lupus*), with dog as *Canus lupus familiaris*.

As the phenotype of the animal is affected by the genes it carries, it is not surprising that there is a relationship between the genetics and physical appearances of most species. The creation and maintenance of a species (speciation) are beyond the scope of this chapter; however, in general terms, when populations become isolated for a long period of time, chance mutations in the DNA result in genetic variations. These alterations to the DNA will only be shared and perpetuated by the members of the isolated group, and hence over time a new form, or species, may result.

4.3   Previous Work in Species Identification

If gross morphological characteristics are present (full or large intact portions of remains), it may simply be a case to identify the remains by microscopy or oesteology (analysis of the skeleton) for species identification so that no further examination is needed. However, if an animal is killed for food or sport, identifying characteristics may be intentionally removed [4], making morphological methods unsuitable [12], and therefore other methods of evaluation are needed [5]. Skins or pelts can be identified from microscopic analysis of the hairs, as the hairs of many animal species have distinct morphological characteristics. However, forensic science laboratories in Asia often get powdered samples with no morphological characteristics for identification [13], rendering this type of analysis unusable.

Other forms of examination, such as scanning electron microscopy (SEM) or gas chromatography coupled with mass spectroscopy (GC-MS), can be used. An inductively coupled plasma mass spectrometer (ICP-MS) can also be used, especially if the area of origin is in question, as captive bred animals may have different concentrations of elements in their tissues as a result of a differing diet from wild types (see chapter 6). Such details may be absent if the remains are as powder or ornaments. In such cases it may be necessary to determine the species by molecular means.

Traditionally, antigen–antibody reactions were used to identify species [14–16] and are indeed still used in many laboratories today [17]. There are several problems with this method of testing. One problem with this technique is that an antibody must first be produced and isolated [14, 15, 18]. This requires a large amount of starting material so that sufficient proteins are extracted for identification [15]. Many proteins lose their biological activity soon after death and can be subject to modification in different cell types [4]. Some antiserum, such as horse (H type), work only in a narrow range of concentrations and, under unstable conditions, can give rise to multiple precipitates [14, 18]. Other commercial antisera often suffer from low titre and cross-reactivity [14, 18]. It may be necessary to titre the batch of antibody prior to use [14].

Since creating an antibody requires a lot of time and effort, and a different antibody is needed for each animal species being tested, new molecular techniques are being developed. In addition, samples that may require analysis include traditional East Asian medicine, bone samples, horns and hairs, sculptures, and powdered remains, which will have insufficient morphology for analysis and may not contain any, or enough, proteins for analysis. However, DNA may be found in all these types of materials. Polymorphisms within this DNA allow for DNA-based techniques to be used to identify species.

4.4   Genetic Polymorphisms

There is a small amount of DNA variation even within a species, known as intraspecies variation (see box 4.1 for comments on DNA). It is estimated that the size of the human genome is about 3.2 billion base pairs. Any two humans taken at random are estimated to share at least 99.8% of the 3.2 billion bases. Humans are thought to have separated from their nearest genetic relative, the chimpanzee, about 6 million years ago. The amount of genetic similarity is greater than 98% and even greater at regions of DNA that encode proteins or are gene related. The differences between humans and chimpanzees are termed interspecies variations.

Box 4.1 DNA—The Basics

DNA is a very simplistic molecule essentially consisting of two extremely long chains, each comprising only a series of four bases. These bases are termed A, G, C, and T. The two chains are wrapped around each other with particular pairing so that a G and C are always across from each other, as are A and T—this forms the famous double helix. Each pair is termed a base pair, and the human genome is estimated to be about 3.2 billion base pairs in total. The human genome is not particularly large and other mammalian species have genomes of comparable size. Plants such as some tree species have genomes that are orders of magnitude larger than the human genome. The 3.2 billion base pairs of the human genome are grouped into 23 sections, termed chromosomes. It should be noted that a chimpanzee has 24 chromosomes and the second largest human chromosome is a fusion of two smaller chimpanzee chromosomes. All the chromosomes are located in the central body of a cell termed the nucleus. In the outer area of a cell (the cytoplasm) there exist small bodies (mitochondria) whose function is to produce energy. Mitochondria have their own DNA, being a circular helical chain.

The function of the DNA is to produce proteins and related biological material. The human genome contains about 25,000 genes. Only about 1.5% of the entire genome is related to this type of biological activity. Much of mammalian DNA appears to have little known function and separates the regions that code for a protein. Over time, changes in the DNA sequence can occur. These changes occur slowly, such that closely related organisms will share much of their DNA if they have a common ancestor. Two humans taken at random share about 99.8% of their DNA, with only 1 base pair being different out of every 1,000. When comparing chimpanzee DNA to that of humans, there is approximately 98.8% homology. As human and species that are related genetically more distantly are examined, the degree of homology reduces. The DNA sequence in a gene may be conserved and varies much less than DNA that is non-gene related. The comparison of the DNA sequence of a gene for two members of the same species will be almost exactly the same. If there are any differences, then this is termed intraspecies diversity. The same DNA sequence, when compared between two different species, may show greater variation, termed interspecies variation. Greater interspecies variation may occur when two species are further apart genetically.

It would be expected that two species that have diverged more recently from a common genetic ancestor will share greater amounts of their DNA compared to a species with a more distantly related common ancestor. Dogs are thought to have been domesticated from wolves about 12,000 years ago

[19]. This is a very short space of time in genetic terms, and hence there is a high degree of homology. Wolves and other members of the Canus family show greater DNA sequence diversity.

In order to make use of the sequence diversity for species testing, it is necessary to first isolate and characterize the sections of DNA that are variable. This may almost certainly require some knowledge of the DNA of the species in question. In some cases this knowledge may not be available.

DNA-based tests that require no prior sequence information can be used, such as random amplification of polymorphic DNA (RAPD), amplified fragment length polymorphism (AFLP), and restriction fragment length polymorphism (RFLP). A comparison of these methods is shown in table 4.2. RAPD, AFLP, and RFLP have been applied to identify animal components [20–31]. The problem and limitation with these whole genome techniques are that there are always issues with reproducibility, and when complex mixtures of two or more species are to be detected, interpretation of results may be difficult due to overlapping restriction patterns that may be generated [20]. It is also difficult to produce accurate reference databases for comparison purposes. It is therefore not only impractical but also logically nonsensical to examine the whole genome when developing a DNA-based species test. It is preferable, then, to use the regions of DNA that show no, or little, intraspecies variation but sufficient interspecies variation for identification purposes. To use known genetic loci, there must be prior knowledge of the genome of the animal being studied. This information is often provided from studies compiling phylogenetic trees attempting to link the ancestry of different animal species or to trace the movement and development of new species. Many of these genes are located on the mitochondrial genome.

4.5 Mitochondrial DNA

The DNA regions used in taxonomy and species testing need to show sufficient interspecies variation to be able to distinguish closely related species. Further, the samples are often old, highly degraded, and powdered. For this reason, and reasons described in the following, the DNA regions used for species testing are on the mitochondrial genome.

The mitochondrial genome is a circular loop of DNA housed within the mitochondrion. Mitochondria are present in the cytoplasm of almost all cells. The number of mitochondria varies between cell types, as the role of the mitochondria is to generate chemical energy, adenosine triphosphate (ATP), by the process of oxidative respiration, in which oxygen is converted to carbon dioxide [46]. Muscle cells contain more mitochondria, and hence more mitochondrial DNA, compared to most skin cells.

Table 4.2   A List of Some of the Current Techniques Used for Species Identification. A Brief Description Is Provided as Well as Advantages and Disadvantages of Each Technique.

Technique	
Advantages	Disadvantages

RAPD: Uses short, random oligonucleotide primers to amplify arbitrary segments of DNA to produce a band pattern specific to an individual or species.

Advantages	Disadvantages
• Gives "fingerprint" of entire genome	• Difficult to reproduce results within and between laboratories
• Cheap	• Cannot produce a database
• Fast	• Cannot reliably separate mixtures [24]
• No prior gene sequence knowledge is needed [32]	• Even among the same species variations exist between different samples [33, 34]
	• Quality and quantity of template will affect the band pattern [33, 35]
	• Buffer, dNTPs, and primer concentration all influence band patterns [35, 36]
	• Major bands shift with degraded DNA [33]
	• Efficiency of primers will affect results [36]

RFLP/AFLP: The use of restriction enzymes to break DNA into fragments. Identification is based on the size and pattern of the resulting band patterns.

Advantages	Disadvantages
• Generates species-specific band patterns	• Requires large amounts of DNA if not coupled with PCR
• Can be coupled with universal or species-specific primers and PCR	• Several restriction enzymes are needed
• No prior gene sequence knowledge is needed (although it is beneficial in order to choose which restriction enzymes to use)	• Cannot reliably separate mixtures of restriction patterns [20, 37]
	• Expensive [12]
	• Laborious and time consuming [12]
	• Variation of restriction sites within species [37, 38]
	• Cannot separate closely related species [37, 39]

Sequencing: Amplification of a segment of DNA for which the base sequence is then determined. Has many different names for the same process: barcoding [40–43], short mitochondrial informative regions (SMIRs) [44], and forensically informative nucleotide sequencing (FINS) [4]. Single nucleotide polymorphisms (SNPs) would also be included in this category as they are an analysis of sequence variation of a smaller scale.

Advantages	Disadvantages
• Can analyze a wide range of species if using universally applicable primers	• Cannot separate mixtures
	• Degraded samples may not generate enough sequence data for identification
	• Susceptible to contamination if using universal primers

Table 4.2   A List of Some of the Current Techniques Used for Species Identification. A Brief Description Is Provided as Well as Advantages and Disadvantages of Each Technique (Continued).

Technique	
Advantages	Disadvantages
Species-specific primers: Isolate SNPs specific to a species and primers are designed based on those SNPs. Products will only be obtained for the species for which the primers were designed.	
• Able to separate complex mixtures [45]	• Prior sequence knowledge is needed to design primers
• It is possible to add additional species as needed	• Only those species for which primers have been designed can be identified
• No need for post-PCR digestion or sequencing	
• Can use multiple genes simultaneously	
• If species-specific primers are coupled with labeled universal primers, can be quite cheap	

Reasons for using mitochondrial DNA rather than DNA within the nucleus include:

- Multiple copies: Each mitochondrion contains its own DNA, with many copies of the circular mitochondrial DNA in every cell. It is thought that each mitochondrion contains between 1 and 15, with an average of 4 to 5, copies of the DNA [47] and there are hundreds, sometimes thousands, of mitochondria per cell. The result is that there are many thousands of copies of the mitochondrial DNA in every cell. This compares with only two copies of nuclear DNA.
- Better protection: The mitochondrion also has a strong protein coat that protects the mitochondrial DNA from degradation by bacterial enzymes. This compares to the nuclear envelope that is relatively weak and liable to degradation.
- Higher rate of evolution: DNA alterations (mutations) occur in a number of ways. One of the most common ways by which mutations occur is during DNA replication. An incorrect DNA base may be added; for example, a C is added instead of a G. This creates a single base change, or polymorphism, resulting in a new form. These single base mutations are rare but occur once every 1,200 bases in the human genome. Single nucleotide polymorphisms are created more commonly than are observed as the errors that occur during DNA replication are corrected (proofread) by an enzyme that exists within the nucleus. This error correcting enzyme does not exist in

the mitochondria [48], and hence errors made when the circular loop of DNA replicates are not corrected. The result is that the rate of change, or evolutionary rate, of mitochondrial DNA is about five times greater than nuclear DNA [48]. This is important in species testing, as even species thought to be closely related may in time accumulate differences in the mitochondrial DNA but show little difference in the nuclear DNA.

- Maternal inheritance: A further reason for the use of mitochondrial DNA in species testing, and in forensic science, is its mode of inheritance. Mitochondria exist within the cytoplasm of cells, including the egg cells. Spermatozoa do not normally pass on mitochondria and only pass on their nuclear DNA. The resulting embryo inherits all its mitochondria from its mother [49–51]. The result is that mothers pass on their mitochondrial DNA type to all their offspring, but only the daughters will pass on the mitochondrial DNA to the next generation. Mitochondrial DNA is therefore passed from generation to generation down the maternal line. There are exceptions to this rule, as demonstrated by mussels of the genus *Mytilus*, which have biparental inheritance of mitochondria [52]. Also, unlike nuclear DNA, where there is a shuffling of the chromosomes at every generation, the mitochondrial DNA does not recombine with any other DNA type and remains intact from generation to generation [50, 53].

The role of DNA is to encode protein and RNA molecules, and the mitochondrial DNA is no different. It possesses 37 genes, encoding 22 tRNA molecules, 2 rRNA molecules, and 13 proteins involved in respiration. The number of bases that comprise the human mitochondrial DNA is about 16,570 base pairs [54]. All mammalian mitochondrial DNA is very similar, with the order and position of the genes being the same. The general structure of the mitochondrial DNA is shown in figure 4.1. There is a site on the mitochondrial DNA where the two strands unzip as part of the replication process, and by convention a base in the middle of this area is called base 1. The DNA bases are numbered sequentially, so in humans the base to one side of base number 1 is base 2, and to the other side it is base 16,570 (note some human mitochondrial DNA types are only 16,569 in length).

The region around base number 1, where the DNA unzips, is the only part of the mammalian mitochondrial DNA that does not encode a protein. As this area is non-coding, it is the one area on the mitochondrial DNA that can develop DNA mutations without affecting the role of the DNA. Greater intraspecies variation occurs at this region compared to the coding regions of the mitochondrial DNA. The DNA bases on either side of base number 1, extending to base 576 in one direction and starting from base 16,024, are called

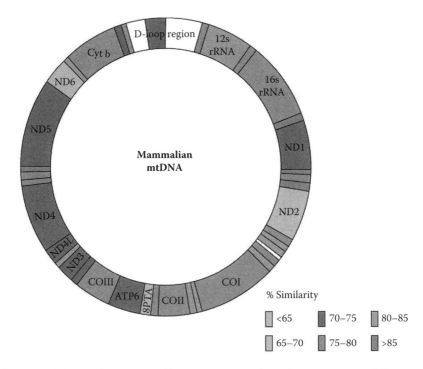

Figure 4.1 (See color insert following page 114.) A schematic view of the mammalian mitochondrial genome. It is a circular loop of double-stranded DNA. Encoded along the loop are 37 genes; these are of various sizes in length and produce proteins or RNA molecules required by the mitochondrion. Base number 1 is at 12:00 as viewed and the bases are numbered sequentially. The bases on either side of base 1 are the only non-coding sections of the DNA, termed hypervariable regions 1 and 2. Next to hypervariable region 1 is the cytochrome *b* gene running from base 14,756 to 15,896 on the reference human mitochondrial genome [54]. The varying colors denote the degree of similarity of the DNA sequences between mammalian species.

the hypervariable regions. These regions, hypervariable region 1 from 16,024 to 16,365 and hypervariable region 2 from 73 to 340, are used in human identification and forensic science to identify a particular individual human. The amount of intraspecies variation precludes their use in species identification.

Gene sequences encode a particular protein, and therefore there is selection pressure against change in the DNA sequence. This conservation of sequence is very important as it preserves the structure and function of the protein produced by the gene. Fewer DNA alterations occur within these encoding regions, resulting in little intraspecies variation. There is also little interspecies variation, as the protein fulfils the same task in a blue whale as it does in a harvest mouse. Proteins are made of amino acids, and amino acids fall into four major categories (acidic, basic, polar, and nonpolar). A mutation can occur that alters, for instance, a basic amino acid for another basic amino acid without a loss of function to the protein. Additionally, a mutation can occur within a gene sequence without any alteration to the amino acid; for instance, the amino acid proline is determined by the three bases CCN, where N is any of the four bases (CCA, CCC, CCG, and CCT, all encode proline). This is an example of how the genetic code is redundant and the third base of the three that encodes an amino acid is often silent. Variation in the DNA sequence of a gene at this position can occur without adversely affecting the protein.

One gene on the mitochondrial DNA in particular has become a tool in species identification, and that gene is called the cytochrome *b* gene.

4.6   Cytochrome *b* Gene

The cytochrome *b* gene is 1,140 bases long and in humans is positioned between 14,756 to 15,896 on the reference human mitochondrial genome [54]. The DNA base number does vary between different mammalian species, but in all cases the cytochrome *b* gene is situated close to one side of hypervariable region 1. It encodes a protein that is 380 amino acids in length [55]; this length is invariable within mammalian species. A schematic diagram showing the position of the cytochrome *b* gene is shown in figure 4.2. Due to the function of the protein, some regions of the amino acid sequence show little variation and other regions have the possibility of greater variation.

Cytochrome *b* is one of 10 proteins that make up complex III of the mitochondrial oxidative phosphorylation system. It is the only one of these proteins encoded by the mitochondrial genome. The protein spans the membrane of the mitochondria and is thought to be involved in electron transfer. Cytochrome *b* crosses the membrane at 8 positions, and these transmembrane spanning amino acid regions show conservation between mammalian species. There is greater variation away from these active regions of the protein. At these less active regions, amino acid variation as described previously can occur. Small variations in the amino sequence encoded by the gene allied with small variation at the third redundant base of some DNA sequences encoding an amino acid lead to variation of the DNA between closely related species.

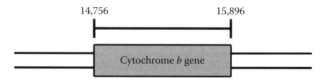

Figure 4.2 A schematic of the cytochrome *b* gene. The gene is close to HV1 on the mitochondrial DNA from 14,756 to 15,896 in humans.

Table 4.3   The Amount of Divergence Using the Entire Cytochrome _b_ Gene for Five Mammalian Species Is Shown. This Is Shown as a Percentage Above the Diagonal and as the Number of Base Variations Below the Diagonal.

	Wolf	Dog	Fox	Cat	Human
Wolf		99.6	84.0	79.0	74.5
Dog	4		83.7	78.9	74.3
Fox	182	186		78.6	73.8
Cat	239	240	244		76.6
Human	291	293	299	267	

Adopted from [56].

The cytochrome _b_ gene was originally used for taxonomic purposes to establish the phylogeny of species [56, 57]. These studies required the comparison of all or part of the DNA sequence encoding the cytochrome _b_ gene. Two members of the same species would be expected to have the same DNA sequence, although small intraspecies variation may result in one or two base differences. Closely related species would be expected to have very similar DNA sequences as they diverged from a common ancestor, and hence the same ancestral species, in the recent past. As the time since a common ancestor increases, it is possible for greater variation to occur, leading to greater differences. If the differences between species are compared, it is possible to determine the amount of variation or similarity. The amount of variation both in DNA bases and as a percentage between the five previously discussed species at the cytochrome _b_ gene is shown in table 4.3.

The degree of homology between the different species mirrors the taxonomic grouping shown in table 4.1. It would be expected that dog and wolf would show very little variation, as it is thought that they have a recent common ancestor (often abbreviated as RCA), more recent than the separation of the common ancestors of the wolf/dog with that of the fox. Figure 4.3 shows a graphical representation of the same five species.

The DNA sequence for the cytochrome _b_ gene is known for a large number of mammalian species. DNA sequences are lodged with the DNA databanks as a repository of genetic information. Two of the largest databanks are the European Molecular Biology Laboratory (EMBL-Bank), maintained at the European Bioinformatics Institute in the UK (www.ebi.ac.uk), and GenBank, maintained by the National Center for Biotechnology Information in the United States (www.ncbi.nih.gov). In the EMBL-Bank alone there are more than 36 billion nucleotide entries from more than 150,000 organisms, and the number is growing continuously. The EMBL-Bank and GenBank joined together with the Japanese-based DNA Databank of Japan (DDBJ), and now information is shared among all three to form the International Nucleotide Sequence Database (INSD).

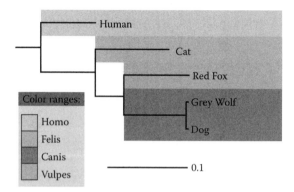

Figure 4.3 (See color insert following page 114.) A phylogenetic tree of the five mammals from table 4.1 analyzed based on the cytochrome *b* difference as described in table 4.2. The scale represents the number of differences between the organisms at the genetic level using the cytochrome *b* gene. The diagram was created using the Interactive Tree of Life [58].

Using the cytochrome *b* gene of the five species as an example (wolf, dog, cat, fox, and human), and when the DNA sequences are compared to the sequences of other closely related and unrelated species, a phylogenetic tree can be produced as in figure 4.4. This tree contains, in majority, mammalian species, but also included are species of bird, fish, and a reptile. The idea behind this diagram is to illustrate that species that are expected to be closely related genetically will group together. A DNA sequence from an unknown sample, when compared to the DNA sequences stored in the DNA databases, will either match perfectly with one of the known samples or will match to a high degree with one of the registered samples.

One problem with the DNA databases is that the DNA sequences registered have not been checked or verified. The accuracy of the sequence data is very much dependent on the supplier registering the data. With relatively recent advances in DNA sequencing there are fewer errors, although this was not the case with the original method of DNA sequencing. Additionally, many nonhuman sequences turn out to be human and are the result of contamination by a laboratory operator. To be confident of the sequence, the source of the original material should be verified.

4.7 Phylogeny Trees

Phylogeny trees are a graphical representation of the relationships between different organisms that share a common ancestor. The phylogeny tree represents in what order these species, or individuals, diverged from a common ancestor. There are many types of data that can be used to construct a phylogenetic tree.

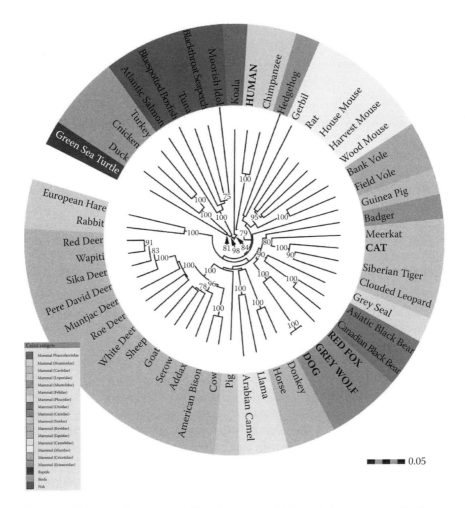

Figure 4.4 (See color insert following page 114.) A phylogeny tree displaying 52 species of animals, the original 5 animals in bold compared to 47 other animals, as produced based on alignment of the cytochrome *b* gene. Clades (groups of organisms with a common ancestor) are shaded to indicate relationships. Expected relationships are observed as dog and wolf, horse and donkey, cat species, and birds and fish are all observed grouped together. Bootstrap values greater than 75 are displayed. The scale represents the number of differences between the organisms in percent. Phylogeny trees and bootstrap values are explained in section 4.4 and box 4.2. Created using the Interactive Tree of Life [58].

For the purposes of this explanation, we will be using nucleotide sequence data from the cytochrome *b* gene on the mitochondrial DNA. For consistency, we will also limit our examples to the five mammalian species (cat, dog, fox, human, and wolf) that we have been using throughout the chapter.

Let us consider the alignment of cat, dog, fox, human, and wolf (figure 4.5). If we were to analyze each sequence against each of the others, we

```
Dog         ATGACCAACATTCGAAAAACCCACCCACTAGCCAAAATTGTTAATAACTCATTCATTGAC
Grey Wolf   ATGACCAACATTCGAAAAACCCACCCACTAGCCAAAATTGTTAATAACTCATTCATTGAC
Red Fox     ATGACCAACATTCGAAAGACTCACCCACTAGCTAAAATCGTAAACGACTCATTCATCGAC
Cat         ATGACCAACATTCGAAAATCACACCCCCTTATCAAAATTATTAATCACTCATTCATCGAT
Human       ATGACCCCAATACGCAAAATTAACCCCCTAATAAAATTAATTAACCACTCATTCATCGAC
            ******    ** ** **     **** **     *** *  * **  ********** **
```

Figure 4.5 A sequence alignment of the first 60 nucleotides of the cyto-chrome b gene on the mitochondrial genome for cat, dog, fox, human, and wolf. Homologous bases between all species are indicated with an * under them.

would find that, according to these sequences, the number of nucleotide dif-ferences between the species can be determined (table 4.4).

This information can then be displayed in a phylogenetic tree, of which there are different types, but the simplest form is the rooted tree. We will use the rooted tree for all of our examples. The root of the tree is the RCA for all of the species shown. From table 4.4 we can see that wolf and dog have less variation than do wolf and fox, which have less variation than wolf and human. From this we can conclude that wolf and dog have a more recent common ancestor than do wolf and fox. This can be converted to several dif-ferent rooted phylogenetic trees (figure 4.6).

The terminal points (represented by the species names) are often referred to as terminal nodes. These represent the extant species data available, whereas the nodes where the data points join (internal nodes) represent a proposed common ancestor. Texts may refer to the extant species as opera-tional taxonomic units. The aim of any phylogenetic tree is to represent the probable genetic distance between two extant species and represent the dis-tance to the most recent common ancestor.

At first glance it appears that all three trees in figure 4.6 are different. In fact, figures 4.6A and B give the same information but are represented in a pictorially different way. The branches in A and B have been scaled according to the genetic distance between the species, and the scales represent the dif-ferences between the sequences in percentages. The only difference between

Table 4.4 The Amount of Divergence Using the Entire Cytochrome *b* Gene for Cat, Dog, Fox, Human, and Wolf Is Shown. The Number of Base Variations Is Shown.

	Wolf	Dog	Fox	Cat	Human
Wolf		4	182	239	291
Dog	4		186	240	293
Fox	182	186		244	299
Cat	239	240	244		267
Human	291	293	299	267	

Adopted from [55].

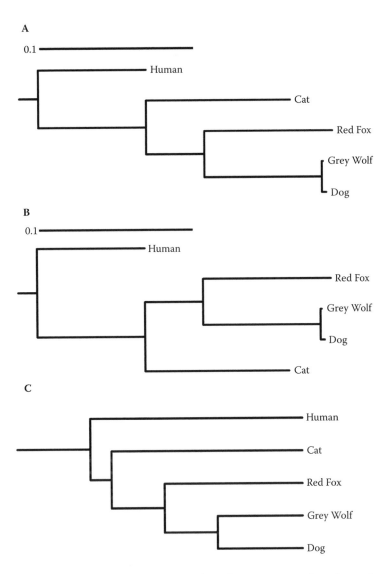

Figure 4.6 Three different rooted trees based on sequence data from the cyto-chrome *b* gene for cat, dog, grey wolf, human, and red fox. A and B are both rooted trees with the lengths of the branches indicating the percentage separation. C is a rooted tree with the branch lengths ignored, which gives less information than do trees A and B. Created using the Interactive Tree of Life [58].

the two is that the branch connecting cat and fox/wolf/dog has been rotated. Branches can be rotated around an axis without altering the table informa-tion. Therefore, red fox and dog/wolf can be rotated, but not human and red fox. Both trees make it obvious that the human sample diverged from the other species at an earlier date, followed by cat, then fox, and finally dog and wolf, which have the most recent common ancestor.

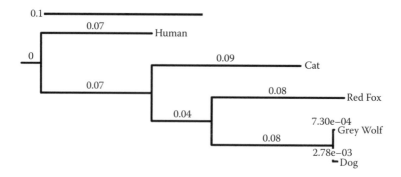

Figure 4.7 A rooted tree based on sequence data from the cytochrome *b* gene for cat, dog, grey wolf, human, and red fox. Branch lengths are displayed. Created using the Interactive Tree of Life [58].

Figure 4.6C, however, gives much less information. All the branches of the tree have been altered to allow the species to line up at the same point, and the figure is therefore nonscaled. This removes much of the information about how much change (how many bases) has occurred between the species, but it still reflects the order of divergence between the species. We can still see that dog and wolf diverged more recently than did cat and human.

The scale represents the percent difference between the species. An example of this is shown in figure 4.7.

It is important to realize that these diagrammatic relationships are based on current DNA data and the RCA is inferred. DNA data can be combined with, or used to support, archaeological remains. Despite the genetic data from sequences on the mitochondrial genome used in species identification and in particular to indicate the degree of diversity between two closely related species, there are few means of confirming whether the degree of genetic relatedness is supported by the data. One test that addresses this problem is called bootstrapping (box 4.2).

Box 4.2 Bootstrapping

Bootstrapping is a standard tool in phylogenetics. It is used to assess the reliability of a phylogenetic tree by taking the aligned DNA sequence data and reshuffling the columns. The reshuffling of the columns allows replacement, so the same column may appear multiple times and a column may not appear as the total number of columns will not change. An example of bootstrapping is shown in figure B4.1.

The idea of bootstrapping was first described by Efron [1] with the idea of re-sampling data many times to determine the degree of confidence in a hypothesis. Since then it has been a standard method in systematics and phylogenetics. It is very relevant to species testing, as a DNA sequence

```
Column number    1  2  3  4  5  6  7  8  9 10 11 12 13 14 15
Dog              A  T  G  A  C  C  A  A  C  A  T  T  C  G  A
Grey Wolf        A  T  G  A  C  C  A  A  C  A  T  T  C  G  A
Red Fox          A  T  G  A  C  C  A  A  C  A  T  T  C  G  A
Cat              A  T  G  A  C  C  A  A  C  A  T  T  C  G  A
Human            A  T  G  A  C  C  C  C  A  A  T  A  C  G  C
                 *  *  *  *  *  *              *  *     *  *


Bootstrap Replicate 1

Column number    2  7 12  8  9  3  4  9 13 15 10  2  6  7  4
Dog              T  A  T  A  C  G  A  C  C  A  A  T  C  A  A
Grey Wolf        T  A  T  A  C  G  A  C  C  A  A  T  C  A  A
Red Fox          T  A  T  A  C  G  A  C  C  A  A  T  C  A  A
Cat              T  A  T  A  C  G  A  C  C  A  A  T  C  A  A
Human            T  C  A  C  A  G  A  A  C  C  A  T  C  C  A
                 *              *  *           *  *  *


Bootstrap Replicate 2

Column number   10  8  6  4 13  3 13  6  2  1  5 10  1 12  7
Dog              A  A  C  A  C  G  C  C  T  A  C  A  A  T  A
Grey Wolf        A  A  C  A  C  G  C  C  T  A  C  A  A  T  A
Red Fox          A  A  C  A  C  G  C  C  T  A  C  A  A  T  A
Cat              A  A  C  A  C  G  C  C  T  A  C  A  A  T  A
Human            A  C  C  A  C  G  C  C  T  A  C  A  A  A  C
                 *     *  *  *  *  *  *  *  *  *  *  *  *  *


Bootstrap Replicate 3

Column number   15  3  4  7 11 12  3  1  2  7  9  8  9  8  5
Dog              A  G  A  A  T  T  G  A  T  A  C  A  C  A  C
Grey Wolf        A  G  A  A  T  T  G  A  T  A  C  A  C  A  C
Red Fox          A  G  A  A  T  T  G  A  T  A  C  A  C  A  C
Cat              A  G  A  A  T  T  G  A  T  A  C  A  C  A  C
Human            C  G  A  C  T  A  G  A  T  C  A  C  A  C  C
                    *  *     *     *  *  *                 *
```

Figure B4.1 How the randomized sequences are produced. The original data set of 15 columns of an aligned DNA sequence is shown. Homologous bases between all species are indicated with an * under them. The same data set is then shuffled based on reordering the columns. The number of columns remain the same (15), but the same column may appear more than once and some columns not at all. This reshuffling is performed between 500 and 1,000 times to obtain a degree of confidence in support of the phylogenetic tree. Real data sets would be the entire DNA sequence and therefore may be 500 bp or more.

from an unknown can be compared to known DNA sequences, aligned, and reshuffled. The higher the bootstrap value, the greater the confidence that the two sequences are homologous. This reshuffling with replacement can be as many as 1,000 times, and each of the reshuffles will generate a phylogenetic tree. If the phylogenetic tree is replicated in all 1,000 reshuffles, then a score of 100% is obtained; but if there is a difference in 20 out of the 1,000 reshuffles, then a score of 98% will be obtained. Ultimately a phylogenetic tree as shown in figure 4.4 is obtained.

Reference

1. Efron, B., Bootstrap methods: another look at the jackknife. *The Annals of Statistics*, 1979, 7(1): 1–26.

4.8   Species Identification Using the Cytochrome *b* Gene

DNA sequences from unknown samples are ultimately compared to the DNA databanks to determine if there is a match to any previously registered DNA sequence. The first step in this identification is the isolation of DNA from the unknown sample.

DNA can be isolated from trace materials with increasing success. Conventional processes work on the basis that the DNA molecule is highly negatively charged and therefore will bind to a substrate that is positively charged. Unbound cellular material can be removed by various solutions, and the DNA is finally isolated from the substrate once almost all the non-DNA cellular components have been removed. Commercial kits are now available that increase the sensitivity and reproducibility of the DNA extraction process.

The second stage is to take the trace amounts of DNA, target the cytochrome *b* region only, and amplify this section of DNA for analysis. The advent of the polymerase chain reaction (PCR) has transformed much of molecular biology, allowing trace amounts of DNA to be routinely analyzed. The PCR process is in essence a high-fidelity photocopier that will copy a specific region of DNA millions of times. The specificity of the copying process is reliant on DNA primers that bind either side of the DNA region to the amplified region. Currently, PCR products of about 800 bases or less in size are amplified routinely. The complete cytochrome *b* gene is slightly larger than the normal size of PCR products that are targeted, in which case smaller sections of the DNA can be amplified.

As the cytochrome *b* gene DNA sequence is very similar for all mammals, it is possible to produce primers that are universal for all the known mammalian species. The benefit of universal primers is that there is no need for any prior information as to the species present in the unknown sample, as the primers will work on all mammalian DNA.

There are a few universal priming sites that have been developed and found to produce a PCR product from all known mammals. An example is the primer pair developed that amplifies a section at the start of the cytochrome *b* gene that is 402 bases in length [59] optimised by [13]. Although this represents less than a third of the entire cytochrome *b* gene, there is sufficient sequence diversity to distinguish almost all mammalian species.

Current technology permits rapid DNA sequencing of sections of DNA produced by the PCR process. The DNA sequences obtained can be compared to the DNA sequences registered at EMBL-Bank or GenBank. Sequence alignments by user friendly software programs such as Basic Local Alignment Search Tool (BLAST) can be performed. The output from the program ranks the DNA sequence matches in degrees of homology. If there is a perfect match between the unknown DNA sequence and the DNA sequence from a species in the databank with 100% homology, then this will be listed first. Those DNA sequences with lessening amounts of homology can be provided. An example of a section of a sequence alignment is shown in figure 4.8.

Figure 4.8 only shows an alignment over 50 bases rather than the entire cytochrome *b* gene, but even within this small section there is sufficient

```
Indian       CCAACGGAGCATCCATATTCTTCATCTGCCTATTTATTCATGTAGGACGA

Javan        CCAACGGAGCATCCATATTCTTCATCTGCCTATTTATTCATGTAGGACGA

White        CCAACGGAGCATCCATATTCTTTATCTGCCTATTCATCCACGTAGGACGC

Black        CTAACGGAGCATCCATATTTTTTATCTGCCTATTCATCCACATAGGACGC

Sumatran     CCAACGGAGCATCCATATTCTTCATCTGCCTATTTATCCACGTAGGACGA

unknown 1    CTAACGGAGCATCCATATTTTTTATCTGCCTATTCATCCACATAGGACGC

unknown 2    CTAACGGAGCATCCATATTTTTTATCTGCCTATTCATCCACATAGGACGC

Bovine       CCAACGGAGCATCAATGTTTTTCATCTGCTTATTTATGCACGTAGGACGA

             * ******* *** ** ** ** ****** **** ** **  *******
```

Figure 4.8 Sequence alignment of DNA sequences registered in EMBL from a small section of the cytochrome *b* gene. Represented are the five extant species of rhino, DNA from two unknown samples, and a cow as an out group. The unknown samples were from powdered remains sold as traditional East Asian medicines. Both the unknown samples have the same DNA sequence as the black rhino (*Diceros bicornis*). The asterisk denotes a DNA base position homologous to all samples. The bases that are highlighted denote those shared by the unknown with the black rhino only.

sequence diversity to be able to exclude the unknown samples as originating from any rhino species other than the black rhino.

The examples described relate to the analysis of traditional East Asian medicines and illegal trade in body parts. Nonhuman DNA testing can also be applied to more mainstream forensic science. Forensic wildlife DNA has been used in a number of countries for criminal investigative purposes, including Canada, Sweden, Austria, and Portugal. In the first instance, pet cat hair found on a murder victim was able to link a suspect to a murder scene and was a key point of evidence in the conviction of a murderer [60]. In Sweden, six cases involving dog and wolf hairs were reported [61]. In all cases it was shown that the suspect hair could not have come from dog hair associated with the suspects except in one case, where there was an inconclusive result [61]. The Austrian case involved correlation of dog hairs found on a murder victim to dog hairs in a suspect's car [62]. The hairs from the victim did not match any of the hairs found in relation to the suspect [62]. In Portugal, it was alleged that a young girl had been murdered and thrown into a pig sty, where small bone fragments were found [63]. After sequencing of the mtDNA, it was concluded that the bone fragments were those of pig (*Sus scrofa*) [63]. In a second case, intestines were found in a forest and thought to be of human origin but did not produce a result with human STR tests [63]. Sequencing of the cytochrome *b* gene was performed and it was concluded that the sample originated from a pig (*Sus scrofa*) [63].

The same principle has been used to identify protected primate species sold as bushmeat, turtle products [64], snake skins [65], and whale meats [66]. The benefit of the test is that even from cooked meats or skins molded into handbags, there is sufficient mitochondrial DNA present to obtain a section of the cytochrome *b* gene. So long as there is a reference sequence in the EMBL-Bank or GenBank, a comparison is possible.

4.8.1   Case Example

An example of extreme sensitivity is that of the test for the presence of the Tibetan antelope (*Pantholops hodgsonii*), otherwise called the chiru [67]. Hairs from this animal are woven into a shawl called a Shahtoosh. This antelope inhabits the high-altitude plains of Tibet and China and produces very fine hair to protect the animal from the environment. The hair from the animal is woven to make shawls of the finest quality that are highly sought after, but it takes between three and five dead antelope to make one shawl. This has led to the catastrophic decline in the numbers of the species, and it is now listed in Appendix I of CITES. Microscopic examination of the hairs taken from the shawls is possible. Hair comparison (chapter 3) plays a crucial role in species identification, although hairs from very closely related species look very similar and it requires skill, and the subjective judgement of the examiner, to be able to determine if a hair from a shawl is a conclusive match to that of the Tibetan antelope. The hairs contain trace amounts of DNA, even after the weaving process in creating the shawl, and these hairs can be recovered and the trace amounts of DNA can be extracted. Fine hairs can be recovered and any cells adhering to the outer surface, such as from human contact, can be removed. Mitochondrial DNA will be present at exceptionally low levels; however, recent advances have permitted the development of DNA-based tests to identify the presence of the Tibetan antelope from hairs taken from shawls. The mitochondrial DNA within the shawl can be isolated and primers used to amplify a small section of the cytochrome *b* gene specific to antelope (see figure 4.9). Under standard conditions even this PCR could not amplify sufficient DNA from the hairs to be detected. A second primer, inside the first, was used to amplify from the PCR products, and a 311 bp fragment, as expected, was produced. This technique, called nested PCR, increases the sensitivity of the test and, if the primers are designed carefully, increases the specificity. The problem with using nested PCR is that with the increase in sensitivity there is a corresponding increase in the chance of contamination of the sample. It is necessary that negative controls are performed and that there is a physical separation of the laboratory where questioned samples are analyzed compared to the positive control voucher specimens.

The 311 base pair fragment was sequenced and found to match that registered in the EMBL-Bank database with 100% homology to *Pantholops hodgsonii*. Cashmere is the nearest material that could be confused with Shahtoosh, but of the 311 bases identified there are 38 differences between the Tibetan antelope and the source of cashmere (*Capra hircus*).

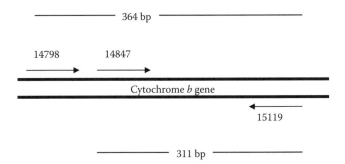

Figure 4.9 The size of PCR products produced by the primers designed to the Tibetan antelope. The outer primers produce a 364 bp fragment that was not detected after the standard conditions. The inner primer amplifies from the first PCR product and is specific to the Tibetan antelope producing a 311 bp fragment. The numbers represent the outermost base number of the primer used based on the human mitochondrial DNA sequence.

4.9 Other Mitochondrial Gene Loci

There are other regions of the mitochondrial DNA genome that have been used in species testing. These include the DNA sequences that encode ribosomal RNA molecules, particularly the 12S rRNA gene. The same methodology is used to examine DNA sequence variation as for the cytochrome *b* gene. As the DNA sequences are highly conserved, it is possible to use universal primers that will amplify a section of the gene. This portion of the gene is sequenced and the DNA sequences are aligned to those registered with the EMBL-Bank or GenBank.

The reason for using 12S rRNA rather than the cytochrome *b* gene is in the case of insufficient sequence variation at the cytochrome *b* gene. Although smaller in size, there are cases where the rRNA gene sequence shows greater variation than the cytochrome *b* gene.

This method of species identification is robust and well validated. It has been used in numerous cases to produce results that are accepted by courts throughout the world. The only issue arises when there is less than 100% homology between the questioned sample and that registered with one of the DNA databases. Due to intraspecies variation, a difference of one base over 400 DNA bases may be acceptable; however, looking at table 4.3, it should be noted that there are only four bases different between dog and wolf over the entire 1,140 bases of the cytochrome *b* gene. Clearly confidence increases with the size of the DNA section that is sequenced.

4.9.1 Bar Code for Life

In an international effort to standardize DNA loci used in species identification, an organization called the Barcode for Life Data System (BOLD) has been established [68]. A section of the cytochrome c oxidase 1 gene (COI) on the mitochondrial genome was proposed. A 648 bp section was considered to have sufficient sequence information as the DNA sequences within this section and can distinguish between 95% of the current species tested. The COI gene is a further locus in the tool of species identification.

A requirement of BOLD is that all samples registered must come from a sample whose species origin can be confirmed if required and that the laboratory providing the sequence DNA works to laboratory standards.

There is much investment in BOLD and a growing repository of DNA in the database based on part of the COI gene. Undoubtedly there will be much value in the use of BOLD as a means of species identification. Ultimately BOLD is a sequence-based species identification tool. The means of species identification described in this chapter relating to cytochrome *b* also apply to other gene loci within the mitochondrial genome.

4.10   Single Nucleotide Polymorphisms

There are many DNA bases that are highly conserved within the cytochrome *b* gene and therefore, although deciphering the entire DNA sequence is possible, there may be only a few signature DNA sequences that need to be examined. When two species are closely related, such as dog and wolf, there may only be four bases that have any significance. It is possible to develop DNA tests that interrogate these individual DNA bases only.

When there is a difference of only one base, termed a single nucleotide polymorphism (SNP), there are a number of methods to detect the DNA base. Whatever method is used, the result looks very different from the sequence data shown previously.

A small part of the DNA alignment of two closely related species, the European beaver (*Castor fiber*) and the Canadian beaver (*Castor canadensis*), is shown in figure 4.10. As can be seen in this figure, the two species share most of the DNA sequence shown.

By using a combination of primers, one of which will bind to both species and the other which will bind to either one species or the other, a species-specific PCR product is obtained. In this case, if DNA from the European beaver is present, a product of 164 bases will be produced, but if Canadian beaver is present, then the product will be 221 bases.

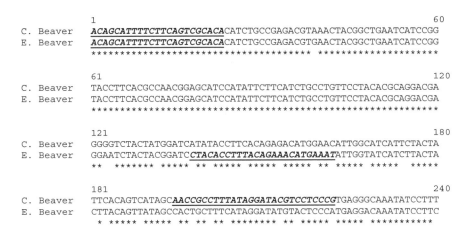

Figure 4.10 Alignment of part of the cytochrome *b* gene for the European beaver and the Canadian beaver. The bases underlined at bases 1–22 act as a PCR primer site for both species. A second primer is made to the European beaver sequence at the base positions 138–164 and is specific for that species and not the Canadian beaver, whereas a primer made to the Canadian beaver sequence at bases 194–221 will bind to the Canadian beaver DNA and not the DNA from the European beaver.

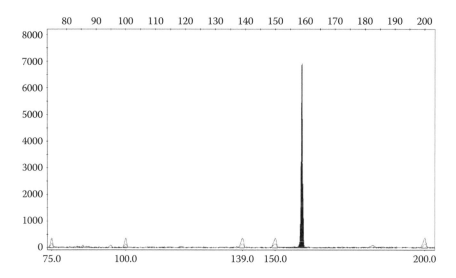

Figure 4.11 A DNA product as the expected position if the unknown was a European beaver. No DNA product was obtained at the position expected for the Canadian beaver.

Using the test shown in figure 4.10, an unknown sample suspected of being beaver can be tested. If beaver is present, only one of the two expected products should be produced. An example of the result is shown in figure 4.11.

Deciphering the full DNA sequence from a section of the mitochondrial genome, such as the cytochrome *b* gene examples, gains the most sequence information. The example of two closely related species shown in the beaver comparison illustrates that within a DNA sequence there may be only a few DNA bases that are informative in separating the two species. These may be a small number of SNPs dotted along the DNA sequence. Using SNP sites can differentiate closely related species, although the amount of sequence information is limited by the number of SNPs detected. In the case above, only one SNP is shown, although more than one SNP site can be examined at any one time. If a number of SNPs for each species are incorporated into the test, then the confidence in the identification of the species will also increase.

4.11   Mixtures

A benefit of using SNP loci is that, unlike DNA sequence information, mixtures of two or more species can be identified. Mixtures of species, particularly when human is one of the species, occur regularly, preventing conventional DNA sequencing generating interpretable results. The use of DNA sequences on the mitochondrial genome permits analysis of DNA from a fraction of one cell. This benefit comes with the problem of inevitable contamination of the

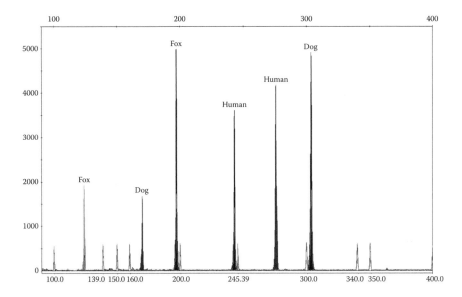

Figure 4.12 A number of colored peaks corresponding to the presence of dog (*Canus lupus familiaris*), European red fox (*Vules vulpes*), and human. The analysis can lead to a prosecution for the illegal killing of a fox with dogs.

sample with traces of human DNA. Cases investigated in the forensic science laboratory include the examination of unknown bloodstains, such as stains thought to be from an animal that has been killed illegally. The removal of a stain from clothing worn next to the skin will result in both the DNA from the white blood cells and skin cells from the wearer of the clothing.

In the investigation of the illegal killing of a fox by dogs, a DNA-based test that can identify the presence of fox and possible contaminating DNA from other animals, such as the dog, was developed. In a way similar to the beaver test, DNA primers were designed to a range of mammalian species including the European fox, the domestic dog, and human DNA sequences [69]. Human primer sets are included to detect the presence of trace levels of human DNA from persons who may have handled the item. The results of the test are shown in figure 4.12, where there is a result for all three species.

It is possible to extend these types of SNP tests further and have the ability to examine numerous species in the one test [69]. Most samples examined in the laboratory will only contain one species plus a human contaminant, but as the nonhuman species is unknown, a test can be used that will produce a characteristic result if at least one of the species is present.

The unlawful killing of animals may be for a number of reasons. This includes poaching for meat such as bushmeat; killing for sport such as bear species; killing for a product such as mementos, clothing including shoes, and bags; and killing for perfumes and supposed medicines. The range of products allows for identification either by microscopy, if possible, or by DNA.

Figure 4.13 An example of a traditional East Asian medicine product. Inside the wrapper were six sheets of plaster as used to apply to the skin. On the reverse of the outer wrapper, written in Mandarin, were the products including tiger, leopard, and musk deer. If any of these species were present, then, firstly, the importation breaks CITES regulations merely by stating that they contain listed species and, secondly, to prove that CITES-listed species are present will require a highly sensitive test.

DNA typing from cooked meat, such as goose products in salami or porpoise from steaks, has been used with much success based on the standard DNA sequencing of part of the mitochondrial DNA.

The identification of rhino from bone and sculptures has been performed as a further example of how the sensitivity and specificity of DNA typing can be applied to wildlife crimes [70].

4.11.1 Case Example

The identification of tiger products in a wide range of possible samples has led to a number of DNA-based tests. An example of this type of testing is in the examination of traditional East Asian medicines, where products are sold as containing tiger products; an example is shown in figure 4.13. In traditional East Asian medicines products there may be a range of animal species, as advertised on the product shown in figure 4.13. The species advertised as present also included leopard, although not stipulated which species of leopard, and musk deer.

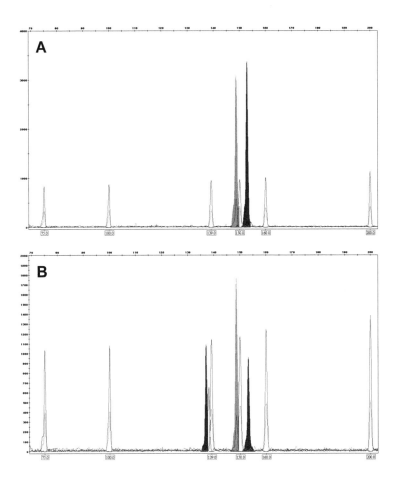

Figure 4.14 Two different samples claiming to contain tiger. (A) Two DNA products corresponding to the expected result if any member of the *Panthera* genus were present in the traditional East Asian medicine. (B) An additional DNA product is present that is produced if a member of the tiger species were present.

A test was developed that would identify all members of the *Panthera* genus, which includes all the tiger species and all the leopard species. This was accomplished by designing DNA primers that would amplify a section of the cytochrome *b* gene only if *Panthera* species are present. A second test was used to amplify a section of the cytochrome *b* gene specific to tiger species. The result of the test is shown in figure 4.14.

The results shown in figure 4.14 indicated that tiger was present in one of the traditional East Asian medicine samples provided and that the other contained a member of the genus *Panthera*. These samples were obtained in Europe and imported from China, making their importation contrary to CITES regulations.

4.12   Problems with Mitochondrial DNA Markers

Part of the benefit of using DNA loci on the mitochondrial DNA is that they are inherited down the maternal line. The benefit is that all members of the population with a common maternal ancestor will have a similar DNA type. The problem comes when hybrid species are produced by the mating of two closely related species, one of which is protected and the other which is not. If the mother is from the nonprotected species, all the offspring will have the same mitochondrial DNA type with no mixture from the protected male species. This problem has occurred within species of sturgeon that produce caviar, some of which are on CITES I. Other closely related species that produce inferior caviar are able to breed with this species but are not offered the same protection. A similar issue arises with many deer species, where the hybrid may not be subject to the legal protection of the original species. In such cases there are DNA markers in the nuclear genome that can be used, although to date less attention has been placed on these loci.

4.13   The Future of Species Testing

There are an increasing number of DNA sequences from a range of different species being lodged with the DNA database. Organizations such as BOLD aid in the standardization of loci used in standard sequence-based methods of species identification. The majority of samples tested will be from a single species, and hence the sequence comparison methods described in this chapter will remain the standard methods of species identification. For those samples where there is a mixture, the method of species identification will change to SNP testing as described in section 4.9.

References

1. Schulz, I., et al., Examination of postmortem animal interference to human remains using cross-species multiplex PCR. *Forensic Science, Medicine, and Pathology*, 2006, 2(2): 95–101.
2. Haglund, W. D., Rodents and human remains, in *Forensic Taphonomy: The Postmortem Fate of Human Remains*, W. D. Haglund and M. H. Sorg, Editors, 1997, CRC Press: London, pp. 405–414.
3. Haglund, W. D., Dogs and coyotes: Postmortem involvement with human remains, in *Forensic Taphonomy: The Postmortem Fate of Human Remains*, W. D. Haglund and M. H. Sorg, Editors, 1997, CRC Press: London, pp. 367–381.
4. Bartlett, S. E. and W. S. Davidson, FINS (Forensically Informative Nucleotide Sequencing)—A procedure for identifying the animal origin of biological specimens. *BioTechniques*, 1992, 12(3): 408–411.

5. Hillier, M. L. and L. S. Bell, Differentiating human bone from animal bone: A review of histological methods. *Journal of Forensic Sciences*, 2007, 52(2): 249–263.

6. Casper, R. M., et al., Detecting prey from DNA in predator scats: A comparison with morphological analysis, using Arctocephalus seals fed a known diet. *Journal of Experimental Marine Biology and Ecology*, 2007, 347(1–2): 144–154.

7. Butler, R., What is a species? *Chemistry and Industry*, 2003, 11: 7.

8. Mallet, J., A species definition for the modern synthesis. *Trends in Ecology & Evolution*, 1995, 10(7): 294–299.

9. Voet, D., J. Voet, and C. Pratt, Life, in *Fundamentals of Biochemistry*, Upgrade Edition, 1999, John Wiley & Sons, Inc.: Brisbane, Chapter 1, pp. 3–21.

10. Mallet, J., Species concepts, in *Evolutionary Genetics: Concepts and Case Studies*, C. Fox and J. Wolf, Editors, 2006, Oxford University Press: New York, pp. 367–373.

11. Olsen, S. J., The fossil ancestry of Canis, in *Origins of the Domestic Dog*, 1985, The University of Arizona Press, Chapter 1, pp. 1–14.

12. Verma, S. K., et al., Was elusive carnivore a panther? DNA typing of faeces reveals the mystery. *Forensic Science International*, 2003, 137(1): 16–20.

13. Hsieh, H.-M., et al., Cytochrome *b* gene for species identification of the conservation animals. *Forensic Science International*, 2001, 122(1): 7–18.

14. Saferstein, R., Identification and grouping of bloodstains, in *Forensic Science Handbook*, 1982, Prentice Hall, Inc.: New Jersey, Chapter 7, pp. 267–296.

15. Balitzki-Korte, B., et al., Species identification by means of pyrosequencing the mitochondrial 12S rRNA gene. *International Journal of Legal Medicine*, 2005, 119(5): 291–294.

16. Prakash, P. S., et al., Mitochondrial 12S rRNA sequence analysis in wildlife forensics. *Current Science*, 2000, 78(10): 1239–1241.

17. Macedo-Silva, A., et al., Hamburger meat identification by dot-ELISA. *Meat Science*, 2000, 56(2): 189–192.

18. Parson, W., et al., Species identification by means of the cytochrome *b* gene. *International Journal of Legal Medicine*, 2000, 114(1): 23–28.

19. Olsen, S. J., Prehistoric dogs in Europe and the Near East, in *Origins of the Domestic Dog*, 1985, The University of Arizona Press, Chapter 6, pp. 71–78.

20. Bottero, M. T., et al., A multiplex polymerase chain reaction for the identification of cows', goats' and sheep's milk in dairy products. *International Dairy Journal*, 2003, 13(4): 277–282.

21. Bravi, C. M., et al., A simple method for domestic animal identification in Argentina using PCR-RFLP analysis of cytochrome *b* gene. *Legal Medicine*, 2004, 6(4): 246–251.

22. Burton, R. S., Molecular tools in marine ecology. *Journal of Experimental Marine Biology and Ecology*, 1996, 200(1–2): 85–101.

23. de los Angeles Barriga-Sosa, I., et al., Inter-specific variation of the mitochondrial r16S gene among silversides, "Peces Blancos" (Atherinopsidae: Menidiinae) and its utilization for species identification. *Aquaculture*, 2005, 250(3–4): 637–651.

24. Guha, S. and V. K. Kashyap, Molecular identification of lizard by RAPD & FINS of mitochondrial 16S rRNA gene. *Legal Medicine*, 2006, 8(1): 5–10.

25. Ishizaki, S., et al., Molecular identification of pufferfish species using PCR amplification and restriction analysis of a segment of the 16S rRNA gene. *Genomics and Proteomics*, 2005, 1(1): 139–144.
26. Partis, L., et al., Evaluation of a DNA fingerprinting method for determining the species origin of meats. *Meat Science*, 2000, 54(4): 369–376.
27. Partis, L. and R. J. Wells, Identification of fish species using random amplified polymorphic DNA (RAPD). *Molecular and Cellular Probes*, 1996, 10(6): 435–441.
28. Sasazaki, S., et al., Development of breed identification markers derived from AFLP in beef cattle. *Meat Science*, 2004, 67(2): 275–280.
29. Sunnucks, P., Efficient genetic markers for population biology. *Trends in Ecology & Evolution*, 2000, 15(5): 199–203.
30. Verkaar, E. L. C., et al., Differentiation of cattle species in beef by PCR-RFLP of mitochondrial and satellite DNA. *Meat Science*, 2002, 60(4): 365–369.
31. Woolfe, M. and S. Primrose, Food forensics: Using DNA technology to combat misdescription and fraud. *Trends in Biotechnology*, 2004, 22(5): 222–226.
32. Shankaranarayanan, P., et al., Genetic variation in Asiatic lions and Indian tigers. *Electrophoresis*, 1997, 18(9): 1693–1700.
33. Lee, J.C.-I. and J.-G. Chang, Random amplified polymorphic DNA polymerase chain reaction (RAPD PCR) fingerprints in forensic species identification. *Forensic Science International*, 1994, 67(2): 103–107.
34. Calvo, J. H., R. Osta, and P. Zaragoza, Species-specific amplification for detection of bovine, ovine and caprine cheese. *Milchwissenschaft*, 2002, 57(8): 444–446.
35. Black, W.C., PCR with arbitrary primers: Approach with care. *Insect Molecular Biology*, 1993, 2(1): 1–6.
36. Welsh, J. and M. McClelland, Fingerprinting genomes using PCR with arbitrary primers. *Nucleic Acids Research*, 1990, 18(24): 7213–7218.
37. Zehner, R., S. Zimmermann, and D. Mebs, RFLP and sequence analysis of the cytochrome b gene of selected animals and man: methodology and forensic application. *International Journal of Legal Medicine*, 1998, 111(6): 323–327.
38. Wolf, C., J. Rentsch, and P. Hubner, PCR-RFLP analysis of mitochondrial DNA: A reliable method for species identification. *Journal of Agricultural and Food Chemistry*, 1999, 47(4): 1350–1355.
39. Meyer, R., et al., Polymerase chain reaction–restriction fragment length polymorphism analysis: A simple method for species identification in food. *Journal of AOAC International*, 1995, 78(6): 1542–1551.
40. Ekrem, T., E. Willassen, and E. Stur, A comprehensive DNA sequence library is essential for identification with DNA barcodes. *Molecular Phylogenetics and Evolution*, 2007, 43(2): 530–542.
41. Vences, M., et al., Comparative performance of the 16S rRNA gene in DNA barcoding of amphibians. *Frontiers in Zoology*, 2005, 2(5).
42. Gelinas, B., Raising the bar on DNA coding: U of G's Biodiversity Institute will perform high-volume "bar-coding," *Daily Mercury*, 2007, Guelph, Ontario.
43. Folmer, O., et al., DNA primers for amplification of mitochondrial cytochrome c oxidase subunit I from diverse metazoan invertebrates. *Molecular Marine Biology and Biotechnology*, 1994, 3: 294–299.
44. Pereira, F., et al., Analysis of inter-specific mitochondrial DNA diversity for accurate species identification. *International Congress Series*, 2006, 1288: 103–105.

45. Kitano, T., et al., Two universal primer sets for species identification among vertebrates. *International Journal of Legal Medicine*, 2007, 121(5): 423–427.

46. Alberts, B., et al., Introduction to cells, in *Essential Cell Biology: An Introduction to the Molecular Biology of the Cell*, 1998, Garland Publishing, Inc.: London, Chapter 1, pp. 1–36.

47. Satoh, M. and T. Kuroiwa, Organization of multiple nucleoids and DNA molecules in mitochondria of a human cell. *Experimental Cell Research*, 1991, 196(1): 137–140.

48. Gray, M. W., Origin and evolution of mitochondrial DNA. *Annual Review of Cell Biology*, 1989, 5: 25–50.

49. Saferstein, R., DNA: A new forensic science tool, in *Criminalistics: An Introduction to Forensic Science*, 2004, Prentice Hall International (UK) Limited: London, Chapter 13, pp. 353–394.

50. Zhang, D.-X. and G. M. Hewitt, Nuclear integrations: Challenges for mitochondrial DNA markers. *Trends in Ecology & Evolution*, 1996, 11(6): 247–251.

51. Wolstenholme, D. R., Animal mitochondrial DNA: Structure and evolution. *International Review of Cytology*, 1992, 141: 173–216.

52. Zouros, E., et al., An unusual type of mitochondrial DNA inheritance in the blue mussel Mytilus. *Proceedings of the National Academy of Sciences of the United States of America*, 1994, 91(16): 7463–7467.

53. Butler, J. M., Mitochondrial DNA analysis, in *Forensic DNA Typing: Biology, Technology and Genetics of STR Markers*, 2005, Elsevier Academic Press: London, Chapter 10, pp. 241–298.

54. Anderson, S., et al., Sequence and organization of the human mitochondrial genome. *Nature*, 1981, 290(5806): 457–465.

55. Linacre, A., Application of mitochondrial DNA technologies in wildlife investigations—species identification. *Forensic Science Review*, 2006, 18(1): 1–8.

56. Kuwayama, R. and T. Ozawa, Phylogenetic relationships among European Red Deer, Wapiti, and Sika Deer inferred from mitochondrial DNA sequences. *Molecular Phylogenetics and Evolution*, 2000, 15(1): 115–123.

57. Ludt, C.J., et al., Mitochondrial DNA phylogeography of red deer (*Cervus elaphus*). *Molecular Phylogenetics and Evolution*, 2004, 31(3): 1064–1083.

58. Letunic, I. and P. Bork, Interactive Tree Of Life (iTOL): An online tool for phylogenetic tree display and annotation. *Bioinformatics*, 2007, 23(1): 127–128.

59. Pääbo, S. et al., Mitochondrial DNA sequences from a 7000 year old brain. *Nucleic Acid Research*, 1988, 16(20): 9775–9787.

60. Menotti-Raymond, M., V. David, and S. O'Brien, Pet cat hair implicates murder suspect. *Nature*, 1997, 386(6627): 774.

61. Savolainen, P. and J. Lundeberg, Forensic evidence based on mtDNA from dog and wolf hairs. *Journal of Forensic Sciences*, 1999, 44(1): 77–81.

62. Nussbaumer, C. and I. Korschineck, Non-human mtDNA helps to exculpate a suspect in a homicide case. *International Congress Series*, 2006, 1288: 136–138.

63. Caine, L., et al., Species identification by cytochrome b gene: Casework samples. *International Congress Series*, 2006, 1288: 145–147.

64. Hsieh, H.-M., et al., Species identification of *Kachuga tecta* using the cytochrome b gene. *Journal of Forensic Sciences*, 2006, 51(1): 52–56.

65. Wong, K.-L., et al., Application of cytochrome b DNA sequences for the authentication of endangered snake species. *Forensic Science International*, 2004, 139(1): 49–55.
66. Grohmann, L., et al., Whale meat from protected species is still being sold in Japanese markets. *Naturwissenschaften*, 1999, 86(7): 350–351.
67. Lee, J.C.-I., et al., DNA profiling of Shahtoosh. *Electrophoresis*, 2006, 27(17): 3359–3362.
68. Hebert, P. D. N., et al., Biological identifications through DNA barcodes. *Proceedings of the Royal Society B: Biological Sciences*, 2003, 270(1512): 313–321.
69. Tobe, S. S. and A. M. T. Linacre, A multiplex assay to identify 18 European mammal species from mixtures using the mitochondrial cytochrome b gene. *Electrophoresis*, 2008, 29(2): 340–347.
70. Hsieh, H.-M., et al., Species identification of rhinoceros horns using the cytochrome b gene. *Forensic Science International*, 2003, 136(1–3): 1–11.

DNA Profiling Markers in Wildlife Forensic Science

5

ROB OGDEN

Contents

5.1   Introduction

Wildlife forensic investigations often require samples to be matched to one another in order to compare individual evidence items or examine familial relatedness. The DNA techniques used to match wildlife samples follow those used for human forensic genetic analysis and are commonly termed 'DNA profiling' methods. DNA profiling has many applications and can involve several types of genetic markers; however, all the techniques work by exploiting the small proportion of the genetic code that varies between individuals within a species.

The history of DNA profiling in wildlife investigations extends back almost as far as human forensic genetic applications. Following the development of the first human DNA fingerprinting methods by Professor Sir Alec Jeffreys in 1985 (Jeffreys et al., 1985), it was quickly realized that parallel techniques could be applied to other species, such as birds (Burke and Bruford, 1987). This work led to forensic genetic evidence being used to support a successful prosecution relating to the theft of wild hawks in the UK as early as 1991. Since this time, wildlife DNA profiling has developed alongside human DNA profiling and has benefited from the horizontal transfer of new analytical techniques.

DNA profiling is now applied to link trace evidence items to victims of wildlife crime in cases of poaching, theft and animal persecution across a wide range of species. It is also routinely used to verify family relationships as part of investigations into the laundering of wild animals through captive breeding programmes. In addition to individual identification, DNA profiling techniques may be used to match a sample to its geographic origin. This has many potential applications, from investigating the illegal timber trade to enforcing fishing regulations, and is complementary to the use of other forensic techniques such as stable isotope analysis (see chapter 6) for determining where a sample has come from.

In contrast to genetic species identification techniques that can be applied to large groups of organisms, DNA profiling techniques are species-specific, requiring the development of new genetic markers and background data for each species under analysis. This has limited the spread of DNA profiling techniques to certain priority species, and there remains a great deal of research to be undertaken before forensic DNA profiling becomes a routine tool in wildlife crime investigations.

This chapter focuses on wildlife DNA profiling techniques, beginning with an introduction to the types of genetic markers used and how they are developed for forensic applications. We will then go on to examine how individual samples are matched, including an explanation of the key concepts of profile exclusion and match probability. This is followed by a look at how

DNA profiles are used to assess the relationships of different individuals to one another. The subject of geographic origin identification is discussed, together with an assessment of the strengths and weaknesses of DNA profiling for providing forensic evidence to investigations into the source of a particular plant or animal. The chapter ends with a look at the future of wildlife DNA profiling and raises some questions regarding the potential for these techniques to have a real impact on the prosecution and reduction of wildlife crime over the coming years.

5.2 Genetic Markers for Wildlife DNA Profiling

5.2.1 Measuring Genetic Variation

The DNA of any individual is largely shared with other animals or plants of the same species. This is particularly true for *coding DNA* that contains the genetic blueprint for functional genes. Genetic variation results from mutations that occur during DNA replication; however, mutations in coding DNA rarely persist within a population, as they are either corrected within the cell or fail to be transmitted to subsequent generations. By contrast, regions of *non-coding DNA*, which are not closely related to any specific gene, exhibit greater mutation rates. In order to examine genetic differences between individuals it is therefore necessary to analyze such regions of non-coding DNA, where genetic variation is more likely to occur.

Specific regions within the genome where mutations have occurred are used as 'genetic markers'. Different types of markers result from different mutation mechanisms, but for all markers it is possible to describe the genetic variation present in a population and to use this information to characterize and differentiate individuals. Each genetic variant at a marker is known as an 'allele'. The DNA profile of an individual is generated from combining observations of the alleles present at a number of unrelated markers.

Genetic markers are found in all species and in all types of DNA. Mitochondrial and chloroplast DNA contain relatively few non-coding regions and typically exhibit mutation rates resulting in markers that vary among higher taxonomic units, such as genera and species, but which are conserved among individuals within a species (chapter 4). Within the nucleus, non-coding DNA mutates relatively quickly, leading to genetic markers that show variations from one individual to another. Furthermore, in sexually reproductive species, nuclear DNA is present in at least two copies, one inherited from each parent, generating genetic variations between generations and among siblings. These factors mean that the genetic markers used in wildlife DNA profiling are generally found in non-coding nuclear DNA.

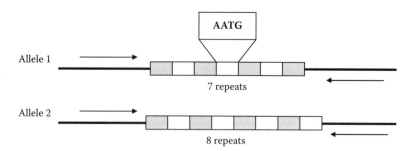

Figure 5.1 Microsatellite markers are comprised of sequences of repeated DNA units 2 to 6 base pairs (bp) long. In this example, the repeat unit (ACAT) is observed seven times in Allele 1 (total length = 7 × 4 = 28 bp) and eight times in Allele 2 (32 bp). The difference in sequence length between alleles is the basis for microsatellite variation.

5.2.2 Microsatellites

The principal type of genetic marker used in DNA profiling is currently the *microsatellite* or short tandem repeat (STR). Microsatellites are regions of non-coding DNA that contain repeated units of between two and six base pairs in length. Variation at microsatellite markers is due to differences in the number of repeat units that alter the length of the DNA sequence (figure 5.1). DNA sequences of different lengths are referred to as separate alleles. In diploid species, such as mammals, two copies of each microsatellite marker are present in every individual. If the two copies are the same length, the individual is said to be *homozygous* for a single allele; if the copies are different lengths, the individual is *heterozygous* for two different alleles.

Microsatellite markers are analyzed by measuring the length of the DNA sequence present in each copy of DNA and thereby identifying the alleles carried by that individual. Microsatellite analysis utilizes the process of PCR (polymerase chain reaction), which allows a genetic marker to be reproduced (or amplified) millions of times. The length of the microsatellite is then measured by comparing the amplified PCR product to a size standard under capillary electrophoresis. In this way, the alleles present in multiple samples are identified and genetic variation among the samples is assessed.

Figure 5.2 shows the analytical results for a microsatellite marker in a family of four birds. The peaks represent the different alleles present in each individual. In this example, each bird is heterozygous, having two different alleles. The two parents have one allele in common, but their other alleles differ. The offspring also have only one allele in common. Offspring 1 has inherited allele B from the father and allele A from the mother; offspring 2 has inherited allele B from the father and C from the mother. The results of this single marker demonstrate four important concepts for DNA profiling: 1) Genetic variation exists between individuals; 2) variation within a family

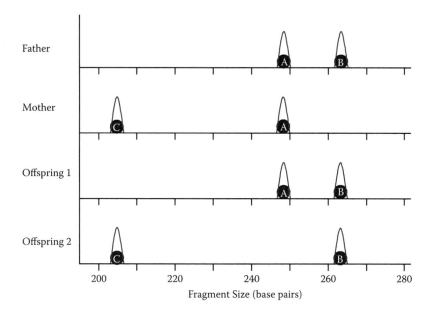

Figure 5.2 The results of microsatellite analysis for a single marker in a family of four goshawks. Each peak represents an allele detected during capillary electrophoresis. A total of three different alleles is observed (A–C). One allele from each parent bird is passed on to the offspring, generating variations between generations and between siblings.

exists between generations and between siblings; 3) alleles are inherited; and 4) it is possible for two individuals to carry the same alleles.

5.2.3 Single Nucleotide Polymorphisms (SNPs)

SNPs are another class of genetic markers that show great potential for wildlife DNA forensic applications. An SNP is described as a specific site in a DNA sequence where a single nucleotide can vary, giving rise to different alleles (figure 5.3). As there are four possible nucleotides at each SNP (A, C, G and T), there are four possible alleles for each SNP marker; however, most SNPs only display two alleles in a population. SNPs occur throughout the genome but, as with microsatellites, SNPs used in sample matching typically occur in non-coding nuclear DNA, where they display bi-parental inheritance and are more likely to exhibit high levels of variation. It is these nuclear SNPs that we refer to when discussing individual sample matching.

SNP markers are analyzed in four principal ways: allele-specific hybridization, primer extension, oligonucleotide ligation and invasive cleavage (Sobrino et al., 2005). In all cases the allele in each copy of DNA is identified, allowing the sample to be characterized at each marker. By comparing the

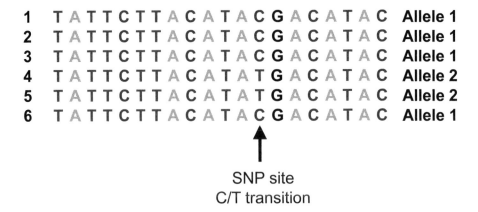

Figure 5.3 An example of a single nucleotide polymorphism (SNP) marker. The six DNA sequences vary at the SNP site between the C and T nucleotides, resulting in two different alleles.

alleles present in different samples, it is possible to begin examining genetic variation or similarity among individuals.

5.2.4   DNA Profiles

Now that we have introduced two of the genetic markers used in DNA profiling, we will turn our attention to how DNA profiles are constructed. A DNA profile consists of a record of the alleles present at a series of genetic markers (figure 5.4). For a single sample, the addition of successive markers decreases the chance of observing that specific combination of alleles in a different

a) Microsatellite profile

Marker	Marker 1		Marker 2		Marker 3		----	Marker 11		Marker 12	
Allele	A1	A2	A1	A2	A1	A2	----	A1	A2	A1	A2
Sample A	11	14	12	18	10	10	----	5	8	9	9
Sample B	11	11	13	17	8	10	----	6	7	9	11
Sample C	12	14	10	11	8	9		5	9	10	12

12 markers x 2 alleles = 24 allele observations

b) SNP profile

Marker	Marker 1		Marker 2		Marker 3		----	Marker 49		Marker 50	
Allele	A1	A2	A1	A2	A1	A2	----	A1	A2	A1	A2
Sample A	1	1	1	2	1	1	----	1	2	1	1
Sample B	1	2	2	2	2	2	----	2	2	1	2
Sample C	1	2	1	1	2	2	----	2	2	1	1

50 markers x 2 alleles = 100 allele observations

Figure 5.4 DNA profiles are produced by combining the alleles observed at multiple markers. Each sample row represents a DNA profile; alleles are numbered separately for each marker. (a) Microsatellite markers exhibit high allelic diversity, resulting in good discrimination among samples. (b) SNP markers are much less variable, requiring more markers to provide the same level of profile discrimination.

individual, until a point occurs where the DNA profile of a sample is statistically very rare within a population. A typical microsatellite profile will list the alleles present at more than a dozen markers, resulting in a profile made up of more than 24 data points; an SNP profile may include up to 50 markers (100 data points), due to the fact that each SNP marker has fewer alleles and therefore does not display the same level of variation among individuals.

By comparing two DNA profiles, it is possible to match or differentiate samples. It is important to note that a DNA profile only represents a snapshot of the entire DNA in a sample; therefore, while different profiles can be said to come from different individuals, matching profiles only indicate the genetic similarity between individuals at those markers in the profile. The significance of a profile match is interpreted by calculating the probability that two individuals in a population have the same profile. At very low probabilities it may be reasonable to conclude that the samples are likely to originate from a single source (or organism). This issue is dealt with in detail in section 5.4 of this chapter.

5.2.5   Wildlife Research vs. Wildlife Forensic Science— Validation of DNA Profiling Markers

The types of markers used in wildlife DNA profiling are the same as those used in a wide range of evolutionary and population genetic studies. Fields such as molecular ecology and conservation genetics generate and apply microsatellite and SNP markers to a wide range of species, and the availability of these markers is potentially very useful to wildlife forensic scientists. However, while the basic tools and techniques are similar, the levels of quality control and accuracy differ between the forensic and research arenas, requiring markers used for DNA profiling to undergo specific validation prior to forensic application.

Validation is the process by which a method is demonstrated to consistently produce reliable, accurate results under a range of experimental conditions. In the case of genetic markers, validation is principally concerned with demonstrating that alleles can be unambiguously identified, that markers exhibit Mendelian inheritance and are independent of one another, and that variation in analytical conditions does not affect the DNA profile produced. As the forensic application of wildlife DNA profiling systems may result in criminal prosecution, the validation standards should parallel those used for human DNA profiling systems; however, this is not always the case. The comparative lack of resources for wildlife forensic research, combined with limited sample availability for many species, often prevents full validation guidelines from being strictly followed. However, it is vital that wildlife geneticists who transfer their techniques to forensic investigation address the issue of validation as fully as possible (Dawnay et al., 2008).

Table 5.1 Sample Types Used to Provide DNA Evidence in Wildlife Forensic Investigations. The Utility of a Sample Type Depends on Its Age and Ambient Environmental Conditions as Well as the Quantity of DNA Present.

Sample Type	Relative Quantity of DNA	Longevity of DNA in a Temperate Environment	Ease of Recovery (Easy–Hard: 1–5)	Nuclear DNA Profiling Ability	Mitochondrial DNA Typing Ability
Tissue	V. High	Weeks	1	Yes	Yes
Blood	High	Weeks	1	Yes	Yes
Skin (dried)	Medium	Years	2	Yes	Yes
Hair (with root)	Medium	Months	2	Yes	Yes
Hair (no root)	Low	Months	3	Limited	Yes
Feather	Medium	Years	2	Yes	Yes
Horn	Low	Years	4	Limited	Yes
Bone	Low	Years	4	Limited	Yes
Faeces	V. Low	Days	5	Limited	Yes
Urine	V. Low	Days	5	Limited	Yes
Fresh leaf tissue	High	Weeks	2	Yes	Yes
Trunk wood	Low	Years	4	Limited	Yes

5. 3   Types of Biological Evidence and DNA Recovery

The successful recovery of DNA from biological evidence is the most important stage in any forensic genetic investigation, and it has therefore been the subject of intense research. DNA is present in every cell of every species; therefore, it should be possible to analyze any type of biological tissue in wildlife forensic genetics. However, when we consider the diversity of animal and plant tissues, from fish scales to feathers, rice grains to tropical timber, it becomes clear that a range of DNA extraction methods needs to be applied to a very broad set of sample materials. This presents the forensic scientist with the challenge of transferring proven DNA extraction methods from human applications, or validating novel extraction techniques to work with unusual samples.

Sample types can be characterized in terms of the quantity of DNA initially present, its protection from environmental degradation and the ease with which purified DNA can be recovered (table 5.1). For example, hard materials such as bone, tooth, horn and ivory may contain relatively little DNA that is difficult to extract but which is preserved in the sample for many years. Animal hairs from the Eurasian badger, *Meles meles*, have been successfully recovered and genetically identified from the vomit and faeces of dogs used in badger persecution; this demonstrates the robust nature of such tissues, as well as the sensitivity of forensic genetic analysis. In contrast, soft tissues tend to contain more DNA that is simple to recover but which is prone to rapid decomposition. Plant tissues vary widely in composition, and certain botanic samples may also contain biochemicals that can inhibit PCR; therefore, different techniques need to be used when dealing with root fibres, leaves, fruit and seeds or solid timber.

One of the particular problems associated with wildlife forensic evidence is environmental degradation. As many crime scenes are outdoors, evidential samples are often exposed to rapid bacterial breakdown, physically destroyed by the elements, or compromised by natural ultraviolet (UV) light that can cause cross-linking of DNA strands, inhibiting subsequent DNA profiling. Another complication to DNA recovery is that crimes against wildlife often involve the illegal trade of processed parts and derivatives, for example, as constituents of certain traditional Asian medicines (chapter 4), and the investigator is often faced with needing to identify heavily treated sample types containing mixed species DNA and PCR inhibitors.

The type and quality of biological evidence obtained during an investigation affect the DNA analysis that can be subsequently applied. As previously discussed, the genetic markers used in DNA profiling are generally found in nuclear DNA and are usually present in two copies per cell (one maternal and one paternal). In contrast, mitochondrial DNA may be present as multiple copies within a mitochondrion, and there may be several thousand mitochondria within a single cell. This results in a much higher success rate for wildlife DNA forensic techniques that rely on mitochondrial markers, such as species identification (chapter 4), and means that DNA profiling techniques really require the presence of multiple cells within an evidence sample to generate reliable results.

5.4   Matching Individual Samples

The purpose of any forensic investigation is to provide evidence that helps piece together what happened during the course of a crime. The ability of forensic genetic analysis to link together the victim, crime scene, suspect or weapon is what has made human DNA profiling such a powerful tool over recent decades. Wildlife crime investigators use DNA profiling to answer the same basic question 'Did this sample come from this individual?'

Forensic genetic methods are usually employed to assess whether trace DNA evidence found at a crime scene or on a suspect originated from the victim of poaching, theft or persecution. For example, poaching investigations may focus on examining whether a sample in the possession of a suspected poacher, such as skin, horn or meat, may have originated from the discarded carcass of an illegally killed animal. In cases of animal persecution, such as badger baiting, individual DNA profiles may be used to link blood samples recovered from clothing, spades, vehicles and dogs to samples recovered from disrupted badger setts, or an individual badger. Such analytical results provide strong evidence for the conviction or exoneration of suspects.

In practice, matching individual wildlife samples is usually a two-stage process, starting with genetic species identification of the trace sample in order to determine the type of evidence available. If the sample originated from a species of relevance for which genetic markers exist, a DNA profile will then be generated. The results of DNA profiling will show one of two outcomes: the profiles differ, in which case the samples are *excluded* from originating from the same source, or the profiles are identical, in which case the statistical probability of the match, or *match probability*, must be calculated.

5.4.1 Profile Exclusion

Determining that two profiles are different and thereby excluding a single sample source is relatively straightforward compared to calculating a match probability, which requires additional DNA profile data from the wider population. For many species of wildlife, obtaining population genetic data is limited by sampling resources; therefore, wildlife investigators need to maximise the use of profile exclusion wherever possible.

The ability of DNA profiling systems to distinguish between two samples can be described as the *power of exclusion*, and this measure can be used to evaluate the relative strength of DNA profiling markers for forensic use. The power of exclusion is a function of the number of alleles observed at each marker, the frequency of each allele in the population and the total number of markers used to generate the DNA profile.

5.4.2 Profile Match Probabilities

When matching DNA profiles are recovered during a wildlife crime investigation, the interpretation of that match is based on assessing the probability that a second individual in the source population has the same profile. Typically, the prosecution will argue that a profile match means that the samples have a single source, while the defence will argue that the two profiles are identical by chance and actually originate from different individuals. The frequency of the DNA profile in the population is therefore the key factor in determining the strength of the evidence provided by a profile match. This section deals with the calculation and presentation of profile match probabilities.

The probability of observing any two alleles at a single marker can be derived from the Hardy–Weinberg equation, which describes the frequencies of homozygous and heterozygous individuals, or genotypes, in a population:

$$p^2 + 2pq + q^2 = 1 \qquad (5.1)$$

where p and q are the frequencies of the two alleles, P and Q, in the population, for a marker where only two alleles are observed (i.e., where $p + q = 1$):

	Maternal allele P p	Maternal allele Q q
Paternal allele P p	pp	pq
Paternal allele Q q	pq	qq

From this equation, it is possible to calculate the probability of observing homozygous and heterozygous genotypes:

$$\text{Prob. (homozygote)} = \Pr(PP) = p^2 \ \text{OR} = \Pr(QQ) = q^2 \tag{5.2}$$

$$\text{Prob. (heterozygote)} = \Pr(PQ) = 2pq \tag{5.3}$$

While most markers have many more than only two variants (alleles), the equation for the homozygote and the heterozygote remains the same. Further, if there are many more alleles than two, the relative number of heterozygotes will increase proportionately to the number of homozygotes.

Table 5.2 shows a punet square to illustrate how combinations of alleles work within a perfect population. If a population is in Hardy–Weinberg equilibrium (HWE), then essentially every generation will result in the same combination as before, and hence the genotype proportions will not change. This is not normally the case, but a large, freely breeding population can be close to a perfect population and be considered in HWE. The simple equations 5.2 and 5.3 can be used as a foundation for determining the genotype frequency.

If the alleles in the population are present at reasonably equal frequencies, then we would expect to see a high number of heterozygote genotypes; conversely, if one allele were very rare and the other very common, the population would contain fewer heterozygotes and tend to be dominated by homozygotes of the common allele. Genetic markers exhibiting high levels of

Table 5.2 Alleles from Mother across the Top and Those from the Father Along the Side to Illustrate That the Theoretical Combinations Will Always Produce Two Heterozygotes for Any Combination and One Homozygote.

	P	Q	R	S	T	U
P	PP	PQ	PR	PS	PT	PU
Q	PQ	QQ	QR	QS	QT	QU
R	PR	QR	RR	RS	RT	RU
S	PS	QS	RS	SS	ST	SU
T	PT	QT	RT	ST	TT	TU
U	PU	QU	RU	SU	TU	UU

Table 5.3   Calculating the Frequency of a DNA Profile Consisting of Three Goshawk Microsatellite Markers (Data from Dawnay et al., 2008). The Genotype Frequency of Each Marker Is Calculated Using Either Equation 5.2 or 5.3, Depending on the Alleles Observed in the Profile. The Profile Frequency Is the Product of the Three Genotype Frequencies. Table from Butler, 2005.

DNA Profile		Population Allele Frequency			Genotype Frequency for Marker		
Marker	Alleles	Times Allele Observed	Size of Database	Allele Frequency	Formula	Frequency	Combined Frequency
Age 5	11	54	100	0.55	$2pq$	0.418	0.418
	14	38		0.38			
Age 7	12	5	100	0.05	$2pq$	0.040	0.017
	18	40		0.40			
Age 9	10	13	100	0.13	p^2	0.017	0.00028
					Profile frequency = 0.00028 (1 in 3539)		

heterozygosity are more powerful at discriminating between individuals and are therefore preferentially selected for DNA profiling.

As we have seen in section 5.2, DNA profiles are constructed from multiple markers. The frequency of a DNA profile sampled from the population is calculated by multiplying the frequencies of the genotypes observed at each marker, following the product rule for independent statistical events (table 5.3). In this way, the profile frequency decreases with increasing marker number, making it less likely that two matching profiles came from different individuals.

It is important to note that the product rule can only be applied to multiple markers if they are known to be independent of one another. Markers that are situated close together on the same chromosome will tend to behave as a single heritable unit and are said to be in linkage disequilibrium (LD). Human genetic markers have been mapped to specific chromosomal locations; therefore, LD can be easily assessed. For most wildlife genetic markers, there are insufficient genomic data to physically map markers, and LD must therefore be investigated using population data.

5.4.3   The Effects of Population Structure

The use of the Hardy–Weinberg equations to calculate the probability of observing a profile in a population assumes that the population itself is in HWE. In reality, natural populations are generally not in HWE, as they tend to violate several of its underlying assumptions, including that of random mating. In any population there are individuals who share DNA with another member of the population because it has been inherited from a common ancestor. In small populations where there is a high degree of inbreeding, the chance that any two members of the population have the same

DNA increases. The degree of inbreeding needs to be taken into account and will affect the chance of two individuals having the same DNA by chance. The equations for calculating the probability of observing homozygous and heterozygous genotypes are therefore adjusted to include a correction for population substructure, as follows:

$$\text{P(hom)} \ (=p^{2}) = \frac{[p(1-\theta)+2\theta][p(1-\theta)+3\theta]}{(1+\theta)(1+2\theta)} \qquad (5.4)$$

$$\text{P(het)} \ (=2pq) = \frac{2[p(1-\theta)+\theta][q(1-\theta)+\theta]}{(1+\theta)(1+2\theta)} \qquad (5.5)$$

where θ represents a measure of population subdivision, usually the estimated value of F_{ST}. Equations 5.4 and 5.5 are commonly referred to as the match probability equations for unrelated samples and were formulated in the 1990s by David Balding and Richard Nichols (1994) for use in human DNA profiling. For a detailed explanation of how these formulae were derived, the reader is referred to Evett and Weir (1998).

In a highly structured population, the effects of inbreeding mean that two individuals within a subpopulation have a higher probability of sharing the same profile by chance than two individuals from different subpopulations. Therefore, using allele frequencies from the total population to calculate the match probability for two samples from within a subpopulation results in an underestimate of the match probability and leads to bias in favour of the prosecution. The inclusion of θ corrects for this bias. These equations were developed to adjust the calculation used in human identification. They can be applied to wildlife, but the value of θ depends on the breeding behaviour of the animal and the size of the population. An example of the effect of θ on the value of the match probability for the goshawk profile is shown in table 5.4. As the population substructure increases (greater values of θ), the probability

Table 5.4   The Effect of Increasing the Value of the Population Structure Correction (θ) on the Value of the Match Probability (P), for the Three-Marker Goshawk Profile in Table 5.3. A Marked Increase in P Is Observed When Theta Rises to 0.1.

θ	P
0	0.00028
0.001	0.00030
0.01	0.00044
0.1	0.00400

of a random profile match increases, providing successively weaker support for the hypothesis that the profiles match because they originate from the same individual.

5.4.4   The Importance of θ in Wildlife DNA Profiling

Although these adjustments are now incorporated into human DNA profiling, there is relatively little natural structure to human populations, and where it does exist, separate allele frequency databases are usually available for each subpopulation. This has led to a value of θ = 0.01 being recommended as a conservative (high) estimate, which usually has relatively little impact on the value of the match probability (table 5.4). In contrast, many wildlife species exhibit much stronger population structuring, and the availability of allele frequency databases is more limited. Examples of θ values for different species profiling systems are provided in table 5.5, demonstrating the importance of correcting for substructure when calculating match probabilities between nonhuman DNA profiles.

Table 5.5   Theta (θ) Values Used to Account for Population Substructure in Match Probability Calculations across Different Species. q > 0.1 for All Nonhuman Population Data Sets.

Species	Region/Database	θ Estimate	Reference
Human	US database		Budowle et al., 2001
Homo sapiens	African Am.	0.0006	
	Asian	0.0039	
	Caucasian	-0.0005	
	Hispanic	0.0021	
	Native Am.	0.0282	
	European database	0.0028	Budowle and Chakraborty, 2001
	Conservative estimate	0.01	Budowle et al., 2001
	Very conservative estimate	0.03	NRCII 4.1, 1996
Dog *Canis familiaris*	US database	0.11	Halverson and Basten, 2005
Cat *Felis silvestris*	US database	0.17	Menotti-Raymond et al., 2005
Badger *Meles meles*	UK database	0.12	Dawnay et al., 2008
Goshawk *Accipter gentilis*	UK database	0.18	Dawnay et al., 2009

5.4.5   Accounting for Individual Inbreeding in Wildlife Species

The inclusion of θ in match probability equations accounts for the increased likelihood of a profile match at the subpopulation level, but it does not address the issue of inbreeding at the level of the individual, where familial relatedness may strongly influence the probability of observing identical profiles in different animals. Individual inbreeding, commonly denoted as f, is ignored in human match probability calculations, as levels of inbreeding in human populations are extremely low. Where a relative is identified as an alternative source for an evidence sample, equations for calculating the match probability for that specific level of relatedness (i.e. parent, sibling, cousin, etc.) are employed. However, for wildlife species, levels of natural inbreeding may be much higher, with individuals living in super-families or displaying strongly hierarchical mating systems that can significantly increase the probability of a profile match. At the same time, the actual relationships between wild animals are rarely known.

Several mechanisms have been suggested to account for inbreeding. The inclusion of f in the match probability equations has been proposed as a correction for inbreeding (Ayres and Overall, 1999). This adjustment effectively increases the match probability for markers with homozygous genotypes and conversely decreases the match probability for heterozygous genotypes. As inbreeding increases the proportion of homozygotes in a population, this is a logical form of correction. However, the effect of this adjustment on a profile for which more than ~50% of markers are heterozygous is to *decrease* the match probability, even when the individual may have many close relatives within the sub-population. A much more conservative method of accounting for f has been proposed by Waits et al. (2001), who suggested that the probability of a profile match should be calculated based on the assumption that all members of the population are related at the level of siblings. This significantly increases the match probability and severely reduces the statistical power of the DNA profiling system. Research into a more balanced solution to account for individual inbreeding in wildlife systems is ongoing.

5.5   The Use of Parentage Analysis in Wildlife Forensic Analysis

The use of DNA to confirm or refute parentage is a well-known concept through its application to paternity testing. The same basic techniques can be applied to wildlife crime investigations that focus on the issue of legally bred versus illegally taken wildlife. Captive breeding programmes are now commonplace throughout the world and focus on either sustaining global populations of highly endangered species or the production of animals and

Table 5.6   Example of Parentage Verification in a Putative Goshawk Family. The Alleles in the Profile of the First Offspring (O1) Correspond to Alleles Present in the Parents at Each Marker, but the Second Offspring is Excluded, as the Three Alleles in Bold are Not Present in the Parents.

	Marker 1		Marker 2		Marker 3		Marker 4	
Bird Profile	Allele1	Allele2	Allele1	Allele2	Allele1	Allele2	Allele1	Allele2
P1	14	16	5	7	12	12	9	10
P2	10	14	6	8	12	13	6	10
O1	14	16	5	6	12	12	6	9
O2	10	16	6	7	**10**	12	7	**8**

plants of commercial value. Problems arise when the two drivers cross over, as they often do, and rare species become highly prized commodities with large profits to be made from their commercial trade. This situation may lead to unscrupulous 'breeders' laundering animals or plants taken from the wild and re-selling them as captive bred individuals. Examples of current issues include the trade in parrots, birds of prey, tortoises and orchids (see box 5.1). Even when commercial breeding programmes for endangered species are completely legitimate, conservationists argue that their very existence supports a market that provides incentive for the illegal capture and sale of wildlife. The suggestion that controlled breeding programmes could supply tiger products to the Chinese medicinal market is one example of such controversy.

The fact that genetic markers are inherited from one generation to the next allows DNA profiles to be used to verify parent–offspring relationships. The alleles present in the DNA profile of an individual must also be present in its putative parents, one allele per marker in each parent. If alleles are observed that do not correspond to those in the putative parental profiles, then the individual can be excluded from being their offspring. The example in table 5.6 extends the analysis of the goshawk family presented earlier. For the first offspring (O1), we can see that for each successive marker, the alleles in the O1 profile correspond to those in the parental profiles. However, for offspring 2, it can be seen that one allele differs at marker 3 and both alleles differ at marker 4, indicating that O2 has inherited these alleles from different birds and is therefore not the offspring of the putative parents.

Box 5.1 Peregrine Family Testing

In 2004 the UK government funded the development of DNA profiling systems for six species of bird of prey, including the peregrine falcon, *Falco peregrinus*. The primary purpose of this research was to improve the forensic genetic methods available for identifying wild birds that had been stolen from nests and laundered though the captive bred bird trade.

Sets of microsatellite markers were assessed and validated for each species, and allele frequency data were subsequently generated for the UK (Dawnay et al., 2009).

One of the first applications of the system was to investigate a falconer who was suspected of providing false parentage records for several peregrines at his breeding centre. DNA samples were obtained from several chicks and their registered parents in order to generate DNA profiles for each bird (table 5.7).

When the profiles were grouped into putative families, it was clear that the chick in Family 1 had a profile consistent with being the true offspring of the adult birds, with every pair of alleles in the chick profile observed in the parent profiles. Based on the population data, the probability of exclusion, PE, was calculated as 0.99968. This is the probability that two birds in the population would be excluded if they were not the true parents. The result therefore provided strong support for the authenticity of the registered family.

The results for Family 2 were very different, with the adult birds excluded as genuine parents at 7 of the 11 microsatellite markers employed (table 5.7). This was sufficient evidence to demonstrate that the chick did not originate from the registered adult pair, raising questions over its true origin. The finding had potentially serious legal implications for the breeder; however, following the initial sample collection, he subsequently admitted that an error in record-keeping may have occurred and produced two more birds that he claimed were the true parents in Family 2. DNA profiles for these birds were generated and compared with the chick (Family 2b). This time all the alleles in the chick profile corresponded to alleles in each of the two parents, and once again the probability of exclusion was sufficiently small to strongly support this second family (2b) as an authentic group.

This method of profile *exclusion* to refute parentage claims, like individual sample matching, does not require profile data from the wider population and is therefore relatively simple to apply. However, the results of parental exclusion are not as definitive as those for individual profile exclusion and require more interpretation. The basis of variability at a genetic marker is the occurrence of heritable mutation events, where one allele changes to another. Although such events rarely occur, they do create the possibility that disagreement between parent and offspring profiles may be due to mutation rather than false parentage. To account for this possibility, the mutation rate at each marker should be incorporated into the interpretation of near-matching profiles.

Table 5.7 DNA Profile Results for Two Registered Families of Peregrine Falcon (Markers Labelled 1–11, Alleles Labelled A1 and A2). Alleles Which are Consistent with True Parentage are in Bold; Alleles Observed in the Chick, but Not the Adults, are in Boxes.

Family	Bird	DNA Profiles (Results observed at eleven microsatellite markers)																					
		1		2		3		4		5		6		7		8		9		10		11	
		A1	A2	A1	A2	A1	A2	A1	A2	A1	A2	A1	A2	A1	A2	A1	A2	A1	A2	A1	A2	A1	A2
1	Male Adult	**B**	C	A	C	A	**B**	A	**B**	A	C	A	C	C	C	**B**	**B**	A	**B**	A	**B**	**B**	K
	Female Adult	C	C	A	**B**	A	**B**	**B**	**B**	A	D	C	C	**B**	D	A	**B**	A	A	A	**B**	**B**	G
	Chick 1	**B**	C	A	A	**B**	**B**	A	**B**	A	D	C	C	**B**	C	A	**B**	A	A	**B**	**B**	**B**	G
2	Male Parent	**B**	C	A	C	A	**B**	A	**B**	A	C	A	C	C	C	**B**	**B**	A	**B**	A	**B**	**B**	K
	Female Parent	A	**B**	C	C	A	A	A	A	A	C	A	A	A	**B**	**B**	**B**	A	A	A	**B**	D	G
	Chick 2	**B**	**B**	[B]	C	A	**B**	A	**B**	[D]	[E]	[B]	C	**B**	[E]	[A]	[A]	[B]	**B**	A	**B**	[A]	[H]
2b	Male Parent	**B**	**B**	**B**	C	**B**	**B**	**B**	**B**	**D**	B	**B**	**B**	B	**E**	**A**	A	**B**	C	A	**B**	**A**	I
	Female Parent	C	**B**	**B**	C	A	A	A	A	C	**E**	C	C	**B**	F	A	**B**	A	**B**	**B**	**B**	H	J
	Chick 2	**B**	**B**	**B**	C	A	**B**	A	**B**	**D**	**E**	**B**	C	**B**	**E**	**A**	**A**	**B**	**B**	A	**B**	**A**	**H**

Mutation rates are marker-specific and can be calculated by observing the frequency of mutations in large pedigree databases. Rates for human DNA profiling markers have been determined in this way; however, this is generally impractical for wildlife species and a conservative estimate of 10^{-3} mutations per generation for all microsatellite markers is therefore recommended (Dawnay et al., 2008). In practice, this means that for profiles comprised of a dozen microsatellites, parental exclusion should be based on differences at more than two loci.

When DNA profile analysis shows that the offspring is consistent with the putative parents, for example, Chick 1 in table 5.7, the statistical significance of the match is evaluated, through the calculation of either a likelihood ratio (LR) or a probability of exclusion (PE). The likelihood ratio evaluates the relative probabilities of claimed parentage versus alternative parentage, based on the genotypes observed in the three profiles. In contrast, the PE is not profile-specific, but represents the combined probabilities of all parental genotypes in the population that can be excluded from producing an offspring with the observed profile. As with match probabilities, the PE is calculated per marker and then multiplied across markers; therefore, profiles with many markers generate powerful PE values and parentage claims can be strongly supported (box 5.1).

5.6   Presenting Matching DNA Profile Evidence

The role of the forensic scientist is to present evidence for the benefit of the court. Although the forensic scientist may be requested to undertake analysis by the prosecution or defence, that analysis must be absolutely objective and presented without bias to either party. In order to maintain a balanced scientific approach, forensic geneticists erect separate hypotheses representing the position of the prosecution and defence. The forensic evidence is then evaluated with respect to these positions. As an expert witness, the forensic scientist is therefore required to comment on the likelihood of the evidence, given each of the alternative hypotheses. He should not express his opinion regarding the likelihood of the hypotheses themselves, e.g., the guilt or innocence of the defendant. This approach has implications for the ways in which DNA profile data are evaluated and presented.

When matching DNA profiles in a wildlife crime investigation, it is conventional to use *Bayes' theorem* to help evaluate the relative likelihood of the two opposing hypotheses, H_p, the prosecution hypothesis, and H_d, the defence hypothesis. A brief overview of how Bayes' theorem is applied to interpret match probability data is provided here; for a more comprehensive explanation, the reader is referred to Evett and Weir (1998).

Bayes' theorem can be applied when considering how the statistical odds for the occurrence of an event may be affected by the subsequent addition of new information. This is of great use when incorporating DNA evidence into an evaluation of the odds associated with a criminal investigation where there is uncertainty surrounding the actual events. The odds form of Bayes' theorem is

$$\text{Posterior odds} = \text{Likelihood ratio} \times \text{Prior odds} \qquad (5.6)$$

This can be written in the context of DNA profile evidence as

$$\frac{P(H_p|E,I)}{P(H_d|E,I)} = \frac{P(E|H_pI)}{P(E|H_dI)} \times \frac{P(H_p|I)}{P(H_d|I)} \qquad (5.7)$$

where the first term is the posterior odds, or the probability of H_p relative to H_d, given the DNA evidence, E, and other circumstances of the case, I. The posterior odds are what the court evaluates at the end of trial. The second term is the likelihood ratio, or the ratio of probabilities of the DNA evidence under either hypothesis; this is what the forensic investigator must evaluate. The third term describes the odds *prior* to the inclusion of the DNA evidence.

These equations appear to be complex, but if considered as words rather than equations, they make sense. The likelihood ratio asks, 'What is the probability of obtaining the DNA result if the prosecution allegation is true compared to the probability of getting the same DNA result if the defence position is true?' Either the prosecution case is true or the defence case is true, but they cannot both be true. Hence, these two statements are mutually exclusive and mutually exhaustive. The prior odds are not prior in time but consider all the other nonscientific evidence such as opportunity, geographical location, age, and gender. This information can help support the prosecution or the defence (or neither).

Calculating the likelihood ratio (LR) is therefore a critical stage in forensic DNA profiling. In the case of matching two DNA profiles, this is achieved by separating the evidence, E, into the two DNA profiles observed, the crime scene (victim) profile, G_C, and the trace evidence profile connected to the suspect, G_S. This gives

$$LR = \frac{P(G_c,G_s|H_p,I)}{P(G_c,G_s|H_d,I)} \qquad (5.8)$$

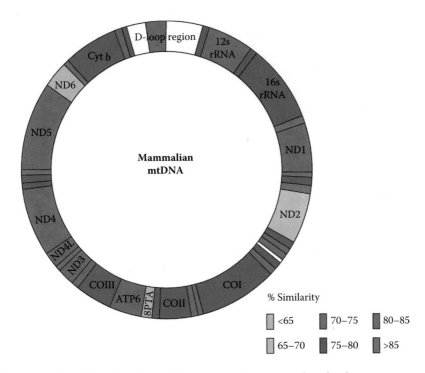

Figure 4.1 A schematic view of the mammalian mitochondrial genome.

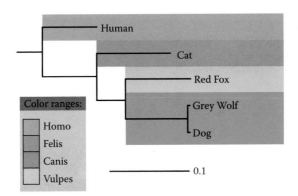

Figure 4.3 A phylogenetic tree of the five mammals from table 4.1 analyzed based on the cytochrome *b* difference as described in table 4.2.

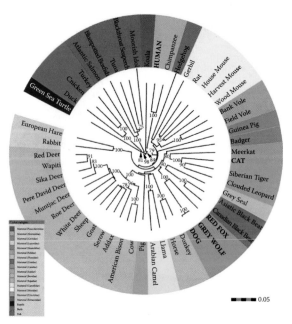

Figure 4.4 A phylogeny tree displaying 52 species of animals, the original 5 animals in bold compared to 47 other animals, as produced based on alignment of the cytochrome *b* gene.

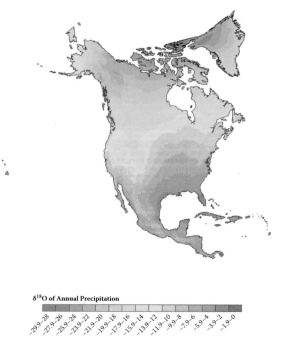

Figure 6.3 An isotope abundance map of North America for the δ¹⁸O isotope in annual precipitation.

which expands to

$$LR = \frac{P(G_s \mid G_c, H_p, I)}{P(G_s \mid G_c, H_d, I)} \times \frac{P(G_c \mid H_p, I)}{P(G_c \mid H_d, I)} \quad (5.9)$$

The right-hand term concerns the probability of observing the victim profile, G_C, in the population. This is not conditional on either hypothesis; therefore, this term can be cancelled out:

$$LR = \frac{P(G_s \mid G_c, H_p, I)}{P(G_s \mid G_c, H_d, I)} \quad (5.10)$$

In the case of matching profiles ($G_S = G_C$), the top line of equation 5.10 represents the probability of observing the trace evidence profile, G_S, if it originated from the victim with profile G_C, which is logically equal to one, giving

$$LR = \frac{1}{P(G_s \mid G_c, H_d, I)} \quad (5.11)$$

The bottom line represents the probability of observing the trace evidence profile, G_S, if it originated from another individual in the population. If we assume that the source of the trace evidence is unrelated to the victim, G_C, then this term (G_C) can be eliminated, to leave $P(G_S \mid H_d, I)$, which can be equated to the match probability. Therefore, the likelihood ratio that allows the forensic investigator to evaluate the probability of the evidence under the prosecution and defence hypotheses is

$$LR = \frac{1}{\text{Match probability}} \quad (5.12)$$

In a practical example, consider an investigation in which a bear has been poached and skinned. Several months later, a bear skin is offered for sale on the Internet. Wildlife investigators obtain a sample of the skin, which the seller claims was legally hunted the previous year, and submit the sample for DNA profiling. The profile from the skin matches that of the poached bear. Based on allele frequencies for the regional bear population, the match probability for the profile is calculated as 3.2×10^{-6}. The prosecution hypothesizes that the bear skin originated from the dead bear (H_p), and the defence hypothesizes that the skin originated from another bear, which has since been destroyed (H_d).

The DNA evidence is evaluated by calculating the likelihood ratio (equation 5.10):

$$LR = \frac{P(G_s \mid G_c, H_p, I)}{P(G_s \mid G_c, H_d, I)}$$

The probability of observing the evidence profile assuming H_p is true is equivalent to the probability that two samples from the same bear have the same profile, which is equal to one. The probability of the evidence assuming H_d is true is the probability that a second bear in the population has the same profile as that already observed in the dead animal, which for unrelated animals is equal to the match probability (3.2×10^{-6}). Therefore, the likelihood ratio becomes

$$= \frac{1}{3.2 \times 10^{-6}}$$

$$= 312,500$$

That is, the evidence is 312,500 times more likely under H_p than under H_d, providing very strong support that the skin originated from the poached bear.

5.7   Geographic Origin Identification

Wildlife legislation usually operates within political boundaries such as national and regional borders or marine fishery zones. Species distributions are governed by biological and environmental factors that rarely coincide with such legislation. This mismatch often leads to wildlife crime investigations asking questions concerning the geographic origin of a sample. The Convention on International Trade in Endangered Species of Wild Fauna and Flora (CITES) controls the international cross-border movement of wildlife (see chapter 2). In order to enforce the Convention's regulations, it is often necessary to demonstrate the geographic source of a specific sample, in addition to identifying the species. Similarly, the effective management of marine-protected areas requires methods that enable illegally harvested stocks to be distinguished from those taken legally from elsewhere. Genetic analysis has been employed to infer geographic origin of samples in biological research for many years; however, the transfer of this application to the wildlife forensic arena is in its infancy.

From a forensic genetic perspective, identifying the geographic origin of a sample is equivalent to identifying its *population* of origin. This is difficult because *population* is a very broad term that can cover a large range of

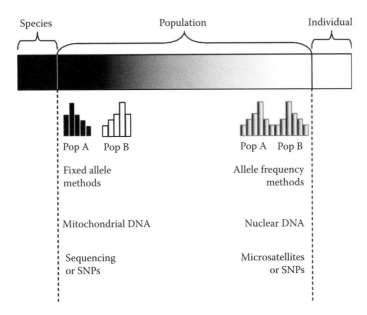

Figure 5.5 Populations identified during geographic origin assignments display genetic variations that range from families to subspecies. Highly divergent populations (to the left) may be identified using fixed alleles via SNP or sequencing techniques to examine mitochondrial DNA. Poorly differentiated populations (to the right) require allele frequency approaches based on nuclear microsatellites or SNPs.

genetic variation, from differences between extended families up to differences between subspecies (figure 5.5). Applying DNA markers that allow us to identify an individual or a species* are well-defined genetic units. In contrast, populations are groups of individuals that breed with one another but can also breed with members of other populations. This sharing of genetic material means that DNA markers are much less likely to show discrete differences among populations, confounding simple, categorical genetic identification.

Geographic origin identification relies on our ability to assign a sample to a particular population, which in turn requires the source population to be sufficiently genetically distinct for DNA markers to differentiate it from other candidate populations. This section examines the forensic genetic methods available for identifying geographic origin across a range of population types, from those that almost deserve the status of separate species to those that are barely recognizable using common genetic markers.

---

* Despite the plethora of species definitions, a species is considered a sufficiently robust category to support specific legislation and subsequent forensic identification.

5.7.1 Highly Divergent Populations—Limited Gene Exchange

Despite being members of the same species, some populations are so isolated from one another that there is effectively no exchange of genetic material between them. Geographic isolation is the most common barrier to gene flow, such as with terrestrial populations on islands, where migration is effectively prevented. In the absence of gene flow, genetic differences gradually build up over evolutionary time to a point where members of an isolated region share the same types of genetic markers within their population but exhibit different marker types to that of any other population. Such markers that exhibit *discrete variation* are very useful for identifying populations and therefore for assigning an individual to a geographic region with a high degree of confidence.

Chapter 4 described how species identification can be achieved through analysis of certain genes found in mitochondrial DNA. In highly divergent populations, mitochondrial DNA (mtDNA) can also provide markers that exhibit discrete differences among populations. The section of mtDNA known as the control region, or D-loop, is a hypervariable stretch of DNA that is often used as a marker in geographic origin identification, with individual control region haplotypes found to be specific to certain populations. A good example of this can be seen in the Chinese sika deer, *Cervus nippon*, which is classified into four sub-species. Following widespread hunting to supply the traditional medicine trade, two sub-species were extirpated from the wild and now exist only in captivity, where they are bred in large numbers for medicinal use. The remaining two subspecies that exist in the wild are critically endangered and heavily protected by Chinese law. In order to enforce this conservation legislation, a method to discriminate one of the wild sub-species from one of the domesticated sub-species was developed, based on mtDNA control region haplotype variation (Wu et al., 2005). Although sequence variation was observed within both sub-species (mean = 0.6% divergence), there was no overlap between them with sequences showing divergences of >5% between sub-species, enabling unknown samples to be confidently assigned to their populations of origin.

From a forensic perspective, discrete differences actually allow us to identify geographic origin through the *exclusion* of all other possible populations. This is an important point, as it follows the established approach for presenting forensic genetic evidence described earlier, requiring the hypotheses concerning geographic origin to be evaluated relative to one another. In the case of discrete mitochondrial haplotypes, this may seem unnecessary; however, as we shall see in the next subsection, the approach is essential when examining genetic evidence for geographic origin among less differentiated populations.

5.7.2   Divergent Populations with Gene Exchange

In the absence of mitochondrial DNA variation, it is necessary to employ genetic markers from the nuclear genome that show variability among regions. Although some microsatellites and SNPs do show discrete differences between populations, individual alleles will often be distributed across populations (figure 5.5). This means that differentiation can only be achieved on the basis of differing *allele frequencies*. The frequency of the alleles observed in a population can be used to characterize its genetic structure and to assess the probability of a sample genotype being observed in that area.

Such approaches have two important implications for wildlife forensic investigation. First, it is necessary to develop large genetic databases to provide representative allele and genotype frequencies for all of the potential source populations. Second, it requires the use of statistical analysis to provide quantitative support for assignment of a sample to each of those populations.

In practice, a DNA profile generated from a test sample will be analysed using one of several population assignment techniques. There is a wide range of software programmes available for assignment, and it is important to understand the statistical basis and underlying assumptions of any that are used. In order to evaluate both prosecution and defence hypotheses, it is essential that the chosen method provides a quantitative estimate of the probability that the sample originated from populations other than the most likely genetic source population. A simple assignment to a single population will not allow a likelihood ratio to be formulated. One commonly used software programme is GeneClass 2 (Piry et al., 2004), which allows the user to examine the support for sample assignment to each of the candidate populations in the data set using a number of assignment methods.

This approach has been used to demonstrate the potential to identify the geographic origin of Chinook salmon for forensic application using SNP markers (box 5.2). In this example, the two populations of salmon were highly divergent due to spatial and temporal differences in spawning behaviours. This resulted in alleles that were almost fixed, displaying large variations in frequency between populations and allowing samples to be identified using relatively few genetic markers. It should be noted that no single statistical method has been agreed upon for forensic genetic population assignment; this is an area of ongoing research and debate (Manel et al., 2005).

Box 5.2 Population Identification in Chinook
Salmon (Schwenke et al., 2006)

Chinook salmon (*Oncorhynchus tshawytscha*) inhabiting the Upper Columbia River basin can be divided into two populations based on geographic separation of their primary spawning grounds and seasonal variation in spawning time (spring-run vs. summer-run). Spring-run salmon are listed under the U.S. Endangered Species Act, while summer-run salmon are not, creating a need to distinguish fish samples from each population for enforcement purposes. To address this issue, three highly informative SNP markers were identified that display strong variations in allele frequencies between two populations. A total of 347 salmon were sampled and profiled using these markers: 98 from the summer-run and 249 from the spring-run. The resulting data were analysed using the programme GeneClass2 to provide a probability of assignment for each of the 18 possible SNP profiles to each of the two populations (table 5.8).

Six SNP profiles were assigned to the spring population and 12 to the summer population. Four of the profiles (1, 2, 7, 8) were excluded from one population or the other; the remaining 14 profiles gave likelihood ratios ranging from over 3000 to just over 1. This wide range in likelihood ratios demonstrates the profile-specific nature of population assignment.

Table 5.8   Results of Population Assignment for Each of the 18 Possible SNP Genotypes. The Likelihood Ratio Describes the Support for the Genetic Evidence Given the Population Assignment, in the Absence of Any Other Prior Information Relating to the Source. Italics Indicate SNP Profiles Not Observed in the 347 Samples.

| | Genotype at Three SNP Markers | | | Probability of Assignment | | |
	1	2	3	Population Assigned	P(summer)	P(spring)	Likelihood Ratio
1	A	BB	BB	Spring	0.0000	1.0000	Exclusion
2	B	BB	BB	Spring	0.0000	1.0000	Exclusion
3	A	AB	BB	Spring	0.0056	0.9944	177.6
4	B	AB	BB	Spring	0.0150	0.9850	65.7
5	A	BB	AB	Spring	0.0154	0.9846	63.9
6	B	BB	AB	Spring	0.0340	0.9660	28.4
7	A	AA	AA	Summer	1.0000	0.0000	Exclusion
8	B	AA	AA	Summer	1.0000	0.0000	Exclusion
9	B	AB	AA	Summer	0.9997	0.0003	3332.3
10	B	AA	AB	Summer	0.9996	0.0004	2499.0
11	A	AB	AA	Summer	0.9967	0.0033	302.0
12	A	AA	AB	Summer	0.9964	0.0036	276.8
13	B	BB	AA	Summer	0.9552	0.0448	21.3
14	B	AA	BB	Summer	0.9145	0.0855	10.7
15	B	AB	AB	Summer	0.9144	0.0856	10.7
16	A	BB	AA	Summer	0.7217	0.2783	2.6
17	A	AB	AB	Summer	0.5659	0.4341	1.3
18	A	AA	BB	Summer	0.5637	0.4363	1.3

The geographic origin of a sample displaying alleles that are very rare in the nonsource population can be identified with a high degree of certainty; samples with alleles common to both populations cannot be reliably assigned.

5.7.3 Populations with High Gene Exchange— Geographic or Morphological Units

In populations that exhibit much larger rates of gene exchange, allele frequencies among different regions may be almost equal. With increasing homogeneity of population gene pools, the number of non-coding markers required to differentiate populations increases to a point where it may be impractical to apply this approach to identify the geographic origin of wildlife samples.

This may occur where a population has been defined by nonbiological criteria, such as fishing zones drawn along geopolitical boundaries or nationally certified timber products obtained from species with wind-blown pollen. In both these examples there is no biological restriction to the dispersal of genes across multiple 'populations' and exact geographic origin cannot be described. Where genetic populations are widely distributed, nongenetic alternatives, such as stable isotopes (chapter 6), may be more suitable to determine the source of biological material.

Other problems occur where the particular biological traits used to characterize a population are controlled by a small number of genes under selection. In this situation, the DNA markers selected from non-coding regions will not display variation that correlates with the biological characters used to describe populations. In order to differentiate populations, it may therefore be necessary to develop genetic markers that target variation in the genes controlling these characters and thereby reflect the population divisions that we have defined rather than the underlying genetic population structure.

5.8 The Future of DNA Profiling in Wildlife Forensic Science

The use of DNA profiling to investigate crimes against wildlife has developed with increasing pace over the past 20 years. With the growth of wildlife forensic science from a specialist pursuit to the early stages of an established field, it may be assumed that the future is very bright for this type of application. However, it is worth remembering that DNA profiling of wildlife is still not considered routine in any country and is not even available in most parts of the world. The species-specific nature of the techniques means that its use has so far been restricted to a few flagship species, and even for these it is often difficult for enforcement officers to secure the funding required to employ DNA profiling during investigations. So what can we expect to see over the next 20 years? The following section examines a few of the possibilities.

5.8.1   Marker Availability

The availability of genetic markers within and across species has been one of the main limitations affecting the development of wildlife DNA profiling. Finding sufficient markers is currently an expensive and time-consuming business, particularly given the marker characteristics that are needed to develop forensically validated profiling systems. Access to large amounts of genomic data has greatly facilitated marker development in humans and other species such as dogs, horses and cattle, for which whole genomes have been sequenced. Until recently it has been unrealistic to expect that large amounts of genomic data would be available to help discover markers in the wide range of species that would benefit from DNA profiling systems. However, with the advent of novel sequencing technologies and the subsequent potential to affordably generate millions of base pairs of DNA sequence per day, it seems likely that we will soon have an enhanced ability to identify genetic markers for many wildlife species.

5.8.2   Marker Type

The use of microsatellite (or STR) markers in DNA profiling appears set to continue for human forensic genetics, despite the ongoing development of parallel SNP-based methods (Butler et al., 2007). Microsatellite markers provide several advantages over SNPs, including their higher power of discrimination, utility in deciphering mixed profiles, ability to be combined in multiplex assays and the fact that they form the basis of large existing population databases. However, wildlife DNA profiling differs from human applications in ways that may support the more widespread development of SNP systems. The need to identify individuals from mixed profiles is encountered a lot less in wildlife crimes, and for most species no population data yet exist, mitigating these potential disadvantages to SNP profiling methods. Geographic origin analysis may also be better addressed using SNP markers that are able to detect adaptive as well as neutral variations. Given the range of potential forensic applications, it therefore seems likely that SNP markers will offer the best solution in the future for at least some species' profiling systems.

5.8.3   Population Data

As we have seen in this chapter, sample matching requires not only a DNA profiling system but also a population database from which to calculate the relevant profile match probability. Generating these background data is likely to pose a significantly greater challenge than developing new marker systems for many species of wildlife. Endangered species are often those subject to illegal activity and are therefore most in need of forensic DNA profiling systems; however, by definition, obtaining sufficient population data for these species is problematic. The potential difficulties and costs associated with producing population data must be considered when proposing the development of DNA profiling systems for rare animals and plants. At the same time, the routine collection of DNA samples that may be used to establish forensic population data needs to be encouraged and coordinated.

5.8.4   Resources

Aside from the technical aspects of wildlife DNA profiling mentioned in the previous sections, the future of these methods is highly dependent on appropriate legislation and resources. Forensic tools are only as useful as the legislation they are employed to enforce, and DNA profiling is no exception. With increasing coordination among responsible agencies and heightened recognition of the value of conserving local species, more and stronger legislation should be developed to facilitate the investigation and prosecution of wildlife crimes. If these efforts are matched by the appropriate resources, then wildlife DNA profiling certainly has the opportunity to make a significant impact, both as an investigative tool and as a deterrent. For wildlife forensic geneticists operating in what is still a narrow niche, this is more likely to be achieved through developing collaborative research programmes and raising awareness of the techniques they offer.

5.9   Summary

- Matching samples using DNA profiling has been applied to wildlife crime investigation for many years, with the techniques developing in parallel to human DNA profiling systems. Applications range from illegal trade to animal persecution.
- DNA profiling is based on the identification of genetic variation that occurs between individuals, families or populations. DNA profiles are produced from the combined measurements of genetic variation at a series of markers.

- Microsatellites and SNPs are the principal types of genetic markers used to create DNA profiles in wildlife forensic genetics. Validation of genetic markers is required prior to use in forensic analysis.
- DNA can be recovered from a wide range of samples; however, the sample type may limit the ability to generate evidential profiles.
- Comparing the DNA profiles of two samples may result in profile exclusion or a profile match; a profile match needs statistical interpretation that requires population genetic data.
- Population substructures and individual inbreeding are likely to have a large effect on profile match probabilities in wildlife species and therefore require careful consideration during forensic analysis.
- DNA profiling is a powerful method for establishing whether an individual has been bred in captivity or potentially taken from the wild.
- Forensic genetic sample matches must be presented with great care and should be evaluated with respect to both the prosecution and defence hypotheses; the use of Bayes' theorem is recommended.
- Matching a sample to a population can allow its geographic origin to be established. The strength of the evidence is dependent on the level of population divergence, which also controls the choice of technique. Statistical models exist for genetic population assignments; however, there is not yet consensus over which technique to use for wildlife forensic applications.
- The future of DNA profiling in wildlife crime investigation is likely to benefit from increased marker availability but may be limited by access to population data for rare or endangered species.

References

Ayres KL and Overall ADJ (1999) Allowing for within-subpopulation inbreeding in forensic match probabilities. *Forensic Science International* 103: 207–216.
Balding DJ and Nichols RA (1994) DNA profile match probability calculation: how to allow for population stratification, relatedness, database selection and single bands. *Forensic Science International* 64: 125–140.
Burke T and Bruford MW (1987) DNA fingerprinting in birds. *Nature* 327: 149–152.
Budowle B and Chakraborty R (2001) Population variation at the CODIS core short tandem repeat loci in Europeans. *Legal Medicine (Tokyo)* 3(1):29–33.
Budowle B, Shea B, Niezgoda S and Chakraborty R (2001) CODIS STR loci data from 41 sample populations. *Journal of Forensic Science* 46(3): 453–489.
Butler JM, Coble MD, Vallone PM (2007) STRs vs. SNPs: thoughts on the future of forensic DNA testing. *Forensic Science, Medicine, and Pathology* 3(3):200-205.

Dawnay N, Ogden R, Thorpe RS, Pope LC, Dawson DA and McEwing R (2008) A forensic STR profiling system for the Eurasian badger: a framework for developing profiling systems for wildlife species. *Forensic Science International: Genetics* 2: 47–53.

Dawnay N, Ogden R, Wetton J, Thorpe RS and McEwing R (2009) Genetic data from 28 STR loci for forensic individual identification and parentage analyses in six bird of prey species. *Forensic Science International: Genetics* (in press).

Evett IW and Weir BS (1998) *Interpreting DNA Evidence: Statistical Genetics for Forensic Scientists*. Sinauer Associates Inc. Sunderland, MA.

Halverson J and Basten C (2005) A PCR multiplex and database for forensic DNA identification of dogs. *Journal of Forensic Sciences* 50: 352–363.

Jeffreys AJ, Wilson V and Thein SL (1985) Hypervariable minisatellite regions in human DNA. *Nature* 314: 67–73.

Manel S, Gaggiotti OE and Waples RS (2005) Assignment methods: matching biological questions with appropriate techniques. *Trends in Ecology and Evolution* 20: 136–142.

Menotti-Raymond MA, David VA, Wachter LL, Butler JM and O'Brien SJ (2005) An STR forensic typing system for genetic individualization of domestic cat (*Felis catus*) samples. *Journal of Forensic Sciences* 50: 1061–1070.

NRCII 4.1—National Research Council of the USA (1996) *The Evaluation of Forensic DNA Evidence*. National Academy Press, Washington, DC.

Piry S, Alapetite A, Cornuet JM, Paetkau D, Baudouin L and Estoup A (2004) Geneclass2: a software for genetic assignment and first-generation migrant detection. *Journal of Heredity* 95: 536–539.

Schwenke PL, Rhydderch JG, Ford MJ, Marshall AR and Park LK (2006) Forensic identification of endangered Chinook Salmon (*Oncorhynchus tshawytscha*) using a multilocus SNP assay. *Conservation Genetics* 7: 983–989.

Sobrino B, Brion M and Carracedo A (2005) SNPs in forensic genetics: a review on SNP typing methodologies. *Forensic Science International* 154: 181–194.

Waits LP, Luikart G and Taberlet P (2001) Estimating the probability of identity among genotypes in natural populations: cautions and guidelines. *Molecular Ecology* 10: 249–256.

Wu H, Wan QH, Fang SG and Zhang SY (2005) Application of mitochondrial DNA sequence analysis in the forensic identification of Chinese sika deer subspecies. *Forensic Science International* 148: 101–105.

Recommended Reading

Balding DJ (2005) *Weight of Evidence for Forensic DNA Profiles*. John Wiley and Sons Ltd., Chichester, UK.

Butler JM (2005) *Forensic DNA Typing: Biology, Technology, and Genetics of STR Markers* (2nd Edition). Elsevier Academic Press, New York,

Evett IW and Weir BS (1998) *Interpreting DNA Evidence: Statistical Genetics for Forensic Scientists*. Sinauer Associates Inc., Sunderland, MA.

Lincoln PJ and Thomson J (1998) *Methods in Molecular Biology, Volume 98: Forensic DNA Profiling Protocols*. Humana Press, Totowa, NJ.

Determining the Geographic Origin of Animal Samples

6

SHANAN S. TOBE

Contents

6.1 Introduction

While the identification of the species from an unknown sample is a first step, and often a crucial one, it may also be important to identify the region of origin of the sample. This becomes an important issue when a species is protected in one region but not in another, or if wild animals are caught and sold as captive bred. The movement of an animal can also be important in cases such as those of stolen animals or even for conservation purposes. One way to determine the origin of a sample is by comparing the ratios of different isotopes using methods such as inductively coupled plasma mass spectrometry (ICP-MS) and isotope ratio mass spectrometry (IRMS). This chapter will give a very basic understanding of the principles of isotopes, ICP-MS and IRMS and their uses in tracing the movements of species. It has been written for the reader with little to no understanding of isotopes or their use in forensic wildlife crime and who comes from a mainly biological background. In addition, case examples using the techniques will be presented.

6.2 Isotopes

To understand the use of ICP-MS and IRMS in wildlife forensic science, and their use in determining the geographic origin of samples, we must first understand isotopes. An atom is composed of a dense nucleus of neutral neutrons and positively charged protons surrounded by negatively charged electrons. The number of protons in an atom is its atomic number, whereas the atomic mass is the total number of protons and neutrons. A chemical element is a type of atom with a given atomic number. In the periodic table of the elements (figure 6.1), chemical elements are classified periodically by their atomic number.

Isotopes are atoms of the same chemical element that have a different number of neutrons in their nucleus. That is, they have the same atomic number but a different atomic mass (figure 6.2). This is recorded as the atomic mass number (n) in superscript followed by the element (X): ^{n}X.

Elements can have two types of isotopes:

- Radioactive isotopes
- Stable isotopes

Radioactive isotopes have nuclei that are unstable and, to regain a more stable form, dissipate energy by radiation. One of the best-known isotopes, the carbon 14 (^{14}C) isotope, is radioactive and decays to nitrogen 14 (^{14}N) at a predictable rate (box 6.1). This radioactivity is the basis for the use of isotopes in many scientific experiments. Radioactive isotopes can be used for radio-isotopic labelling, are substituted into reactions to trace chemical processes and were also used for early DNA analysis. Based on the radioactivity of the isotopes, they can be detected using X-ray film.

Stable isotopes, as the name implies, have nuclei that are stable and therefore do not undergo radioactive decay. All but 12 elements exist in stable isotopic form [1, 2]. Each element that has isotopes has a light isotope that is dominant (present in much higher numbers) and one or more heavy isotopes [3, 4]. The percentage of an element's isotopes in nature is called its isotopic abundance. The heavy isotopes have a natural (isotopic) abundance of a few percent or less [2]. For example, there are three naturally occurring Oxygen isotopes: ^{16}O, ^{17}O and ^{18}O. Their isotopic abundances are, respectively, 99.759 %, 0.037 % and 0.204 % [5]. Heavy isotopes generally react in the same way as their corresponding dominant isotopes, although there are slight differences. Heavier isotopes form stronger bonds than their lighter counterparts and can take longer to react due to their larger masses; this is known as kinetic isotope effects (KIEs), or mass discriminating effects [1, 2].

Thermodynamic isotope effects (TIEs) are another common isotope effect. TIEs relate to the energy state of a system and to the physico-chemical properties of the element being analysed such as infrared absorption, molar volume, boiling point and melting point [1, 2]. The effects are evident in processes where chemical bonds are not formed or broken [2].

In forensic science, for determining the geographic origin of a sample, we use stable isotopes. These isotopes are incorporated into tissues through biological processes and can then be detected in both living and dead organisms.

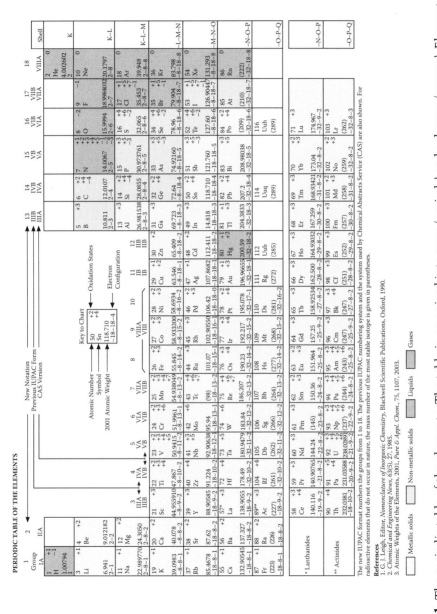

Figure 6.1 The periodic table of the elements. The table is organised into rows (periods) and columns (groups). Elements in the same group share similar characteristics. More information on the periodic table can be found in any general chemistry textbook.

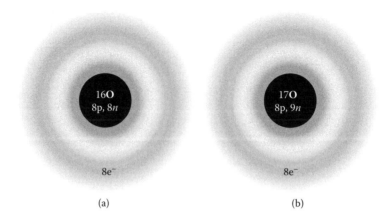

Figure 6.2 Comparison of two oxygen (O) isotopes. (a) A basic atom of ^{16}O containing 8 of each protons (p), neutrons (n) and electrons (e-). (b) A ^{17}O isotope containing 8 protons and electrons, but containing an extra neutron.

In tissues that grow at a constant rate, these isotopes are deposited in a linear fashion as the tissue grows. The isotopes can then be detected using a suitable technique, and a 'timeline' of the larger movements of an animal can be determined.

Box 6.1 Carbon 14 Dating

Carbon is the sixth element in the periodic table and, in its basic state, contains six neutrons and six protons. The number of neutrons can vary to seven (^{13}C) or eight (^{14}C). ^{12}C and ^{13}C are both stable and nonradioactive, but ^{14}C decays to ^{14}N at a predictable rate. The decay rate of ^{14}C is 5720 ± 47 years; this value was determined by Willard Libby and his colleagues in 1949 [1], for which he received the Nobel Prize in Chemistry in 1960 (http://nobelprize.org/nobel_prizes/chemistry/laureates/1960/). It is by using this constant decay rate that scientists are able to determine the age of a biological object, a process called radiocarbon dating or carbon 14 dating. Carbon 14 is so important because C occupies the central position in the chemistry of biological systems and will therefore be incorporated into every biological system [2].

Carbon dating has been in use since its discovery in 1949 and is probably the best-known use of isotopes. It is assumed that ^{14}C is produced at a constant rate and is present in the atmosphere in the range of 1 part per trillion (1 ppt). It can date biological objects up to about 60,000 years ago. ^{14}C is produced in the atmosphere when cosmic radiation collides with an atom and creates an energetic neutron. This energetic neutron can then collide with an atom of N and by knocking out a proton will create an atom of ^{14}C. This C isotope can combine with O to create carbon dioxide (CO_2), which is then absorbed by plants. Through photosynthesis the ^{14}C atom is incorporated into plants, which are subsequently eaten by animals. The ^{14}C is then incorporated into bones for animals and wood for plants. When an organism dies, it is no longer replenishing its amount of ^{14}C and, based on the steady decay rate, an age can be determined for the sample based on the proportion of remaining ^{14}C to ^{12}C.

The equation for the production of ^{14}C is

$$^1n + {}^{14}N \rightarrow {}^{14}C + {}^1H \text{ [3]}$$

The equation for the decay of ^{14}C is

$$^{14}C \rightarrow {}^{14}N + e^- + v_e \text{ [4]}$$

where n is a neutron, e^- is an electron and v_e is an anti-neutrino. See [4] for a detailed review of ^{14}C and its use in radiocarbon dating.

References

1. Engelkemeir, A. G., W. H. Hamill, M. G. Inghram and W. F. Libby, The half-life of radiocarbon (C14). *Physical Review*, 1949, 75(12): 1825–1833.
2. Kamen, M. D., Early history of carbon-14. *Science*, 1963, 140(3567): 584–590.
3. Ruben, S. and M. D. Kamen, Long-lived radioactive carbon: C14. *Physical Review*, 1941, 59(4): 349–354.
4. Bronk Ramsey, C., Radiocarbon dating: Revolutions in understanding. *Archaeometry*, 2008, 50(2): 249–275.

Planetary isotope ratios were determined with the formation of the earth and are generally fixed [1, 2]. Isotopic ratios between species or climatic regions vary constantly as a result of biological, biochemical, chemical and physical processes [1], known as isotope fractionation [4]. It is these isotopic variations that we are interested in as forensic scientists. The isotope ratios in a sample are measured relative to the isotope ratios of standards and are reported as a delta (δ) value. The standards that are used vary depending on which isotope is being analysed, and generally the δ values are arbitrarily set to 0‰ for the standards [2]. C is analysed relative to calcium carbonate, called Vienna Pee Dee Belemnite (VPDB); hydrogen (H) and O are analysed relative to the Vienna Standard Mean Ocean Water (VSMOW) and nitrogen (N) is analysed relative to air.

Variations in stable isotope abundance ratios are typically small and are generally expressed in parts per thousand (per mill or ‰) [4]. To obtain a δ value, the equation used is

$$\delta = \left[\frac{R_{sample}}{R_{standard}} - 1 \right] \times 1000$$

where δ has units of per mill (‰) and R is the ratio of heavy to light isotopes [2, 4, 6].

The relative abundance of one isotope, say ^{18}O, in one geographical area may significantly differ from another area. The distributions of ^{18}O and ^2H are determined by temperature and precipitation—that is, with increasing temperature, precipitation contains more of the heavier isotopes (this fractionation occurs at a rate of approximately 0.5‰ for every °C for O) [7]. In addition, the initial liquid phase of rain (evaporation from the ocean) is enriched with ^{18}O and ^2H compared to later precipitation—a phenomenon known as rainout [7]. A detailed explanation can be found at http://www.sahra.arizona.edu/programs/isotopes/oxygen.html. By sampling areas throughout the world, maps can be created of the isotopic abundance of, for example, ^{18}O (figure 6.3). The maps show the change of the isotope ratio over geographical area relative to VSMOW and has units of ‰. As illustrated in figure 6.3, this change can be gradual. By determining the isotope ratio for ^{18}O, and by using a map such as in figure 6.3, a general area of origin can be determined for a sample.

From these maps one can see that the δ value decreases as the latitude increases, the δ value decreases toward the centre of the continent (due to the rainout effect) and sharp changes are demonstrated in the mountains (toward the west of the continent). Based on these maps, migration patterns, geographical origin and movement of species can be traced using tissue samples. It is important to note, however, that the type of tissue sample obtained can have a major impact on the results, as different tissues have different turnover rates, which will be discussed in section 6.5. The isotopes most commonly used in forensic science for this purpose are generally H, C, N and O, although more can be brought into use as isotopic abundance maps are created and their usefulness is assessed. Some of the rarer elements are also occasionally used. A full list of isotopes can be obtained from the National Institute of Standards and Technology (NIST). In order to separate and detect the different isotopes in a sample, specialist equipment is needed. We will discuss ICP-MS and IRMS and their use in separating and analysing isotopes.

6.3   Inductively Coupled Plasma Mass Spectrometry (ICP-MS)

ICP-MS is a sensitive technique capable of determining a range of metals and nonmetals at concentrations below one part per billion (1 ppb). It is also

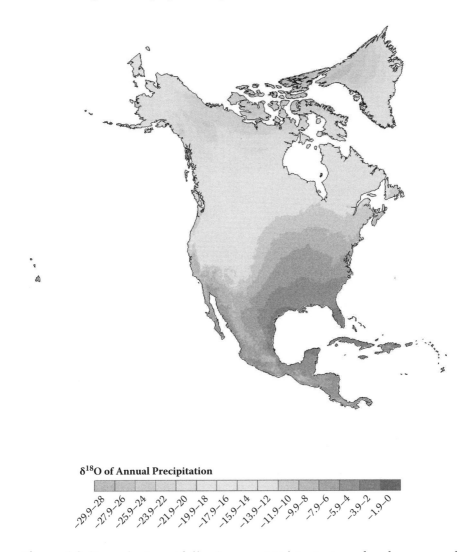

δ^{18}O of Annual Precipitation

-29.9–28 -27.9–26 -25.9–24 -23.9–22 -21.9–20 -19.9–18 -17.9–16 -15.9–14 -13.9–12 -11.9–10 -9.9–8 -7.9–6 -5.9–4 -3.9–2 -1.9–0

Figure 6.3 (See color insert following page 114.) An isotope abundance map of North America for the δ^{18}O isotope in annual precipitation. The map shows the ratio of ^{18}O relative to VSMOW and has units of per mill (‰). This figure is used with kind permission from Dr. Gabriel Bowen, was obtained from http://www.waterisotopes.org and is based on data from [8].

capable of monitoring isotopic speciation for ions of choice [9]. A full description of this technique is beyond the scope of this chapter (entire textbooks are available—see the end of this chapter) and only a brief and simplified version of the operation of this instrument will be described.

In order to isolate and detect the isotopes in any sample, the sample must first be ionized. Using high temperature plasma discharges, ICP-MS generates positive ions, which are then detected on a mass spectrometer. ICP exhibits

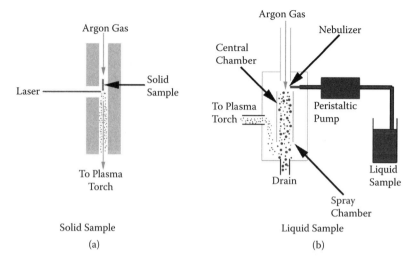

Figure 6.4 Basic examples of how samples are prepared for injection into the plasma torch of the ICP-MS. (a) Solid samples are vaporized using a laser prior to injection. (b) Liquid samples are aerosolized via the nebulizer into small and large droplets. The large droplets are removed in the spray chamber from which they are drained, leaving the small droplets to be injected into the plasma torch.

high plasma (4,500–8,000 K) and electron (8,000–10,000 K) temperatures, a long (2–3 ms) plasma/sample interaction time and a high (1–3 × 10^{15} cm^{-3}) electron density [10]. These conditions result in the complete vaporization and atomization of the sample and reduce the interferences often found in other types of atomic spectrometries [10].

Samples are injected as either a gas, solid or liquid, although the system was designed primarily for liquid samples [11]. Multiple methods are available for injection, but they all result in an aerosol that is then ionized by the plasma [11]. Gas samples are injected directly into the instrument. Solid samples can be analysed using lasers to vaporize the sample prior to injection (figure 6.4a); however, this is still quite rare and suffers from calibration difficulties. Liquid samples, the most common type of sample, are injected as a fine aerosol [11] (figure 6.4b). This is accomplished by the use of a pneumatic nebulizer, to vaporize the sample, followed by a spray chamber to remove droplets that are too large for the instrument before introduction into the plasma torch [12].

The liquid sample is first pumped (at ~1 mL/min) through a peristaltic pump, which pumps the sample into the nebulizer at a constant rate [11]; however, viscous solvent systems should be avoided [13]. Once in the nebulizer, the sample is broken into a fine aerosol using a flow of argon (Ar) gas (at ~1 L/min), similar to when a deodorant is sprayed from a canister [11]. Ar, a noble gas, is used because it is chemically inert and has the capability

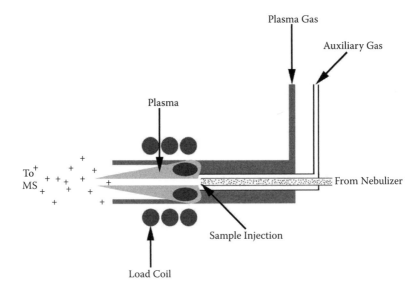

Figure 6.5 The plasma torch. The plasma gas (usually Ar) is passed between the outer and middle tubes. Ions are created when atoms collide with high-energy electrons from the plasma and are directed to the MS.

to excite and ionize most of the periodic table elements [12]. The aerosol contains both small and large particles that would cause a noisy signal if they entered the plasma together. Small, uniformly sized droplets give a steadier signal, so it is beneficial to remove the larger aerosol particles. This is accomplished with the use of a spray chamber whose function is to only allow small particles to pass to the plasma torch and to smooth out any pulses caused by the peristaltic pump [11]. The most common spray chamber, the one we will discuss here and which is illustrated in figure 6.4b, is the double-pass spray chamber [11].

Although there are many designs, the main purpose of the spray chamber is to only allow the small droplets of fluid to pass to the plasma torch. The double-pass spray chamber consists of a chamber with a central tube running nearly the entire length. The aerosol is directed into the central tube and, when it emerges, the larger droplets are removed by gravity and the smaller droplets are forced back into the outer chamber [11]. The small droplets are then directed to the sample injector of the plasma torch [11, 13].

The plasma (a body of gas with high levels, and equal numbers, of ions and free electrons) is formed in a stream of Ar gas flowing through the plasma torch [12] (figure 6.5). There is a copper induction coil at the top of the torch that is connected to a free-running or crystal-controlled radiofrequency (RF) generator [10]. Then a current is passed through the RF generator and a magnetic field is produced by the induction coil, which in turn induces current in the Ar gas, which then forms a plasma [9, 10, 12]. As long as the magnetic field and symmetrical Ar gas flow are maintained, the plasma will be stable and self-sustaining [10].

The plasma torch is made up of three tubes, the inner and outer tubes and the sample injector [12], which are usually made of quartz [13]. Generally, Ar gas is used for the torch and is passed between the outer and middle tubes at a rate of 10 to 15 L/min [13]. Gas also flows between the middle and inner tubes (the auxiliary gas) and in the inner tube (from the nebulizer), both at about 1 L/min [14]. The gas from the nebulizer carries the sample into the plasma in the form of a fine aerosol [15] and also punches a hole through the plasma.

The most important part of the ICP system is the ionisation of the sample for detection with the mass spectrometer. This is accomplished by the different regions of the plasma. The sample is first introduced into the plasma in the gas flow from the nebulizer in the form of a fine aerosol [15]. The first step is to evaporate the water off the sample to leave a very small amount of solid [14]. This solid is then turned into a gas and then into ground state atoms [14] as it passes through the extreme heat of the plasma. The ground state atoms are converted to ions by colliding with energetic electrons from the Ar [9, 14]. The ions then emerge from the plasma and are directed to the mass spectrometer by the interface region.

The interface region is an important part of the ICP-MS; however, for the purposes of this chapter, it is sufficient to say that it transports the ions from the plasma torch (at a pressure of 1 atm) to the mass spectrometer (at a pressure of $\approx -10^{-8}$ atm) [16]. The purpose of the mass analyser is to separate the ions of interest based on their mass-to-charge (m/z) ratio from all of the other ions present (from the Ar, non-analytes, matrix and solvent) [17]. The mass spectrometer can be set to select a single ion based on the m/z ratio or the mass spectrum can be scanned for a complete overview of all m/z ratios [13]. We will discuss the most common type of mass analyser, the quadrupole mass filter technology—although double focusing instruments are also possible [12, 13, 18].

The quadrupole consists of four parallel cylindrical metal rods of the same length and diameter, usually made of stainless steel or molybdenum and sometimes coated in ceramic to resist corrosion [9, 17, 19]. One opposite pair of rods is attached to the positive side of a variable direct current (DC) source and the other pair to the negative side, and variable RF potentials are applied to each pair of rods [18]. By changing the DC-RF voltage, ions of interest are accelerated through the middle of the rods while those ions not of interest are ejected from the quadrupole or spiral into the rods and are converted to neutral molecules [17–19]. This means that only ions with stable trajectories, the ions of interest, are transmitted to the detector [12]. This process can then be repeated at a different DC-RF voltage until all the analytes of interest have been analysed. A schematic example of this process can be seen in figure 6.6.

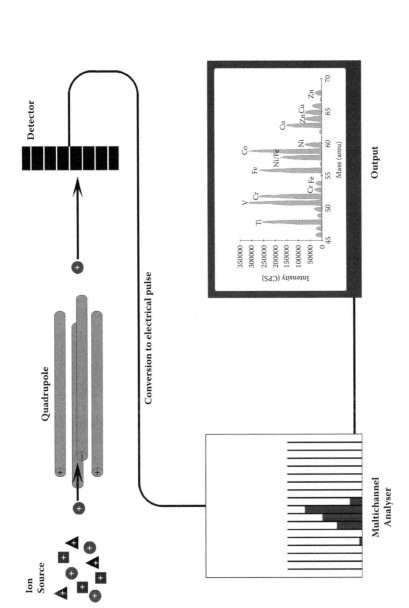

Figure 6.6 The quadrupole mass spectrometer. By selecting a specific DC-RF voltage, an ion of interest will pass through the centre of the quadrupole rods and emerge at the other end at the detector. As the ions are detected, they are converted to an electrical pulse, then separated and stored by the multichannel analyser based on their mass-to-charge ratio (m/z). The DC-RF voltage is then changed to correspond to another ion of interest, and this process is repeated until all of the different elements have been analysed. The output data shown were obtained from a Philips ICP-MS. Copyright S.S. Tobe; used with permission.

After passing through the quadrupole, the ion is converted into an electrical pulse by the detector. The ion of interest is repeatedly scanned for, and the electrical pulses (from the ion of interest) are sorted and counted by a multichannel analyser [17]. As the electrical pulses are counted in each channel, a profile of the mass is built up corresponding to the spectral peak for the ion of interest [17]. In practice, 25 elements can be estimated, with good precision and in duplicate, in 1 to 2 min [17]. The output is displayed as a spectrum, with abundance on the x axis and mass-to-charge ratio (m/z) on the y axis (figure 6.6).

6.4 Isotope Ratio Mass Spectrometry (IRMS)

An IRMS is a specialised type of mass spectrometer that produces precise and accurate measurements of light stable isotopes relative to a standard. It is generally divided into three parts: the ion source, the mass analyser and the ion collection assembly [2]. For IRMS analysis, samples must be converted to a gas that is isotopically representative of the whole sample [2]. The gas is usually introduced into the IRMS via one of two methods: dual inlet (figure 6.7a) or continuous flow. Continuous flow IRMS allows the IRMS to be connected to a range of preparation devices, of which we will discuss two types: bulk stable isotope analysis (BSIA) (figure 6.7b) and compound specific isotope analysis (CSIA) (figure 6.7c). IRMS is used to detect the ratios of $^{13}C/^{12}C$, $^{2}H/^{1}H$, $^{15}N/^{14}N$, $^{18}O/^{16}O$ and $^{34}S/^{32}S$, but as the sample must enter the IRMS as a gas, what are actually analysed are the gases carbon dioxide (CO_2), hydrogen (H_2), nitrogen (N_2), oxygen (O) and sulphur dioxide (SO_2) [2, 20].

In dual inlet IRMS (figure 6.7a) the sample is converted to simple gases offline by an apparatus involving vacuum lines, compression pumps, concentrators, reaction furnaces and micro-distillation equipment [2]. It is the most precise type of IRMS, but it is also time consuming, large sample sizes are required and isotopic fractionation is possible at each stage [1, 2, 20]. Once converted to a simple gas, the sample enters the IRMS by a variable volume gas reservoir, called a bellows [2, 20]. Reference gas is also pumped into the IRMS via a separate bellows to allow comparison of the sample and reference gases under identical conditions [2]. The bellows also equalizes the pressure of the two gasses to equalize flow into the IRMS, so that pressure effects on the ion current ratios are avoided [20]. With this method the relative isotope ratios are determined; that is, the measurement obtained with the instrument reflects the isotope ratio of the sample relative to the standard rather than the absolute isotope ratio of the sample [20].

Whereas duel inlet IRMS offers high precision for stable isotope measurements, the continuous flow IRMS (CF-IRMS) has several advantages. Some of these advantages include online sample preparation, smaller sample sizes,

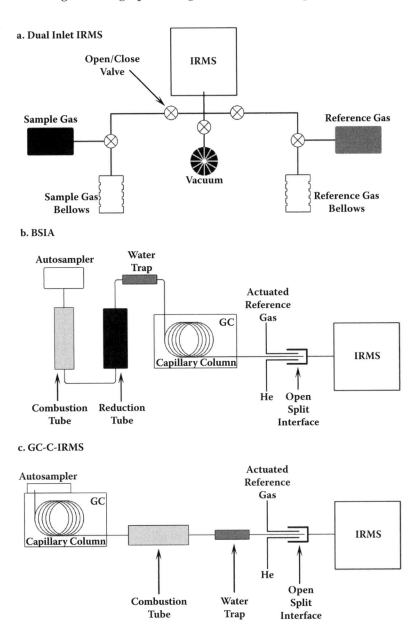

Figure 6.7 Three of the different sample introduction techniques for IRMS. (a) Dual inlet IRMS, which equalizes the pressure of the sample and reference gasses with bellows. (b) Bulk stable isotope analysis (BSIA) with an Elemental Analyser (EA). The analysis of O and H isotope ratios does not have combustion or reduction tubes and instead has only one pyrolysis furnace. (c) Compound specific isotope analysis (CSIA), also known as GC-C-IRMS. The components of the combustion tube differ depending on which element is being analysed.

faster and simpler analysis. Furthermore, CF-IRMS allows the instrument to be interfaced more easily with other preparation techniques like elemental analysis, gas chromatography (GC) or high-pressure liquid chromatography (HPLC) [2]. As a result, CF-IRMS is the technique most used for isotopic analysis in forensic research and casework [2].

The CF-IRMS allows the IRMS to be connected with different automated sample preparation devices and was originally used for the measurement of nitrogen isotopes [1]. CF-IRMS has been expanded to include C, S, O and H [2]. Helium (He) is used as a carrier to sweep the analyte gas into the ion source [1, 20] and to purge the samples to prevent the introduction of H_2O, O_2 and N_2 [2]. The sample is converted into a simple gas and therefore the isotopic values generated are representative of the whole sample (components of mixtures are not separated) in a technique called bulk stable isotope analysis (BSIA) [2] (figure 6.7b). The inlet devices for BSIA are elemental analysers (EAs) of different natures depending on which isotopes are being analysed. With BSIA analysis time can be as fast as 6 min per sample [1, 21].

To analyse N, C and S isotope ratios, a combustion EA is used [2]. The sample is placed in a sealed tin capsule in the autosampler, where it will fall into a combustion tube containing, among other materials, an oxidation catalyst [2] (such as copper [Cu] and chromium oxide for the analysis of C and S [1, 21]). A pulse of O_2 temporarily replaces the He and the sample combusts, with the temperature rising from 1,000°C to 1,700°C, resulting in the production of N_2, NOx, CO_2 and H_2O gases [2, 21]. The products are swept through a reduction tube, at 600°C, where NOx is reduced to N_2 and excess O_2 is removed, and then through a trap to remove H_2O [1, 2, 21]. The remaining gases are then separated on a gas chromatography (GC) column, and directed to the IRMS, where N and C isotope ratios can be measured together [2, 21]. For analysis of S isotope ratios, the oxidation and reduction tubes are packed with a different material [2]. Some of the GC effluent also enters the IRMS through an open split interface via a capillary tube [2]. The capillary tube is the interface between the IRMS and the EA and reduces the gas flow for the IRMS [2].

For the analysis of H and O isotope ratios, a high temperature conversion EA (TC/EA) is used [2]. The sample is placed in a sealed silver capsule in the autosampler, where it will fall into a combustion tube [2]. The sample is pyrolytically decomposed at temperatures that can exceed 1,450 °C [2] and converted to H_2 and CO gasses [1, 21]. These gases are again separated on a GC column before analysis on the IRMS via an open split interface [2]. H and O isotope ratios can be measured simultaneously [2].

In compound specific isotope analysis (CSIA) (figure 6.7c), also known as GC-IRMS and more recently as GC-Combustion-IRMS (GC-C-IRMS), a stream of He is used to drive the analyte through the stages of separation and preparation for the IRMS [20]. The isotopic compositions of individual components are measured from the sample [2]. The sample is first dissolved with an organic solvent prior to injection into the GC [2]. The components in the sample are separated on the GC column, and after then enter a combustion furnace and a reduction tube. They are then dried by a water trap before being analysed by the IRMS [20]. A splitter at the end of the GC column sends most (>95%) of the sample to the combustion tube [2]. Different procedures are required for the analysis of C, N, O and H using GC-C-IRMS. When measuring N or C isotopes, the sample is passed to a combustion furnace, before entering a reduction tube to remove excess O_2 [2, 21]. N measurements require an additional reduction furnace filled with Copper at a temperature of 600°C to convert NOx to elemental N_2 and a liquid nitrogen CO_2 trap to remove ions formed by the CO_2, which interfere with the isotope measurements for N [1, 2, 20, 21]. H and O analysis is prepared by passing organic compounds through a high temperature furnace (>1,000°C, usually with a metal reactant), where H_2 or CO is quantitatively released [2, 20]. The gases are dried on a water trap to remove H_2O and then pass to the IRMS via an open split interface [2].

Once the sample is converted to a gas via either the dual inlet or the continuous flow, it enters the IRMS [2] (figure 6.8). The gas enters in either a pure form or as a band in the He carrier gas across a flow restrictor to maintain a steady operating pressure in the ion source [20]. The gas enters the ionisation chamber and is bombarded by a focused electron beam in a high vacuum. The ion source is designed to contain the neutral gas molecules for as long as possible to maximise the probability of ionisation by the 70 eV electron beam, and it is referred to as 'tight' [20]. This results in the formation of positive ions by electrons from the beam, causing a loss of electrons from the sample molecules [2]. The ions are then repelled out of the ionisation chamber through a flight tube between the poles of an electromagnet, which separates the ions according to their m/z ratio [2, 20]. The trajectory of the ion out of the electromagnetic field will be determined by its m/z.

After the ions leave the magnetic field, they are collected by Faraday cups in the collector array [2, 20]. The Faraday cups are positioned specifically to receive the ions of a given m/z [1]. The MS of an IRMS has sacrificed its flexibility for accurate and precise isotope measurements [1]. While a classic MS possesses a collector able to record mass over a wide range, the IRMS only possesses up to eight collectors (Faraday cups), each tuned to the specific m/z that is to be measured. The Faraday cups are positioned so that all of the different ions coming from the ion source are detected simultaneously, with each ion accounting for one charge [2]. The ion currents are amplified and digitised using a voltage-to-frequency converter and are transferred to a computer [2]. The sensitivity of the IRMS is typically 10^{-3} or better; that is, 1,000 molecules enter the ion source for every ion detected [1]. The computer integrates the peak area, which represents the ion signal, for each isotopomer

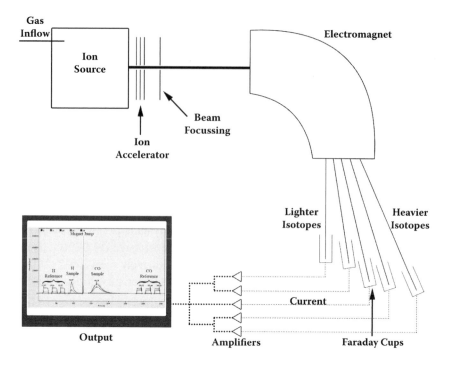

Figure 6.8 A schematic of an IRMS. Sample gas is ionized and ejected. The ion beams are separated by the electromagnet and monitored by the Faraday cup detectors. The ion currents are then amplified and digitised. The output shown is from the analysis of hydrogen and oxygen and is kindly provided by Dr. Nicola Farmer. The first three square peaks are the hydrogen reference gas, the next peak is the H sample peak, the large peak is the magnet jump for the analysis of H to CO, the broad peak is the CO sample peak and the final three square peaks are the CO reference gas.

and calculates the corresponding ratios [2, 20]. Before and after sample analysis, reference gasses with known isotopic values are passed through the IRMS and the sample values are worked out by comparison.

6.5 Uses of Isotopes in Forensic Wildlife Crimes and Conservation

As the use of isotope analysis becomes more commonplace in the scientific community, it is being used increasingly in wildlife forensic science to trace the movements of animals as well as determine the geographical origin of individuals [22]. Previous methods included the analysis of stomach and faecal contents to try to determine the origin of a given animal. This is not practical, as it can only give a 'snapshot' of the organism's most recent meal(s) and would be rendered useless if the animal had been in captivity for several days or weeks, or if the sample was presented as a pelt, powder or other trace sample. Isotope ratio analysis overcomes these problems by utilising samples after the isotopes from food have been incorporated into the organism. As a result it can give information on the organism over a relatively long period of time. Soft tissue can also be analysed to determine the origin of other samples such as bushmeats. The detection of stable isotopes in soft tissue can only be used to determine the location of an animal shortly prior to death, as soft tissues are replenished continuously [22].

It is important to note that the type of tissue and the type of animal from which that tissue originates can have a major impact on the results. Tissue type is important as different types of tissues have different incorporation rates and are replenished at different rates [23]. Tissues with a high isotopic turnover include the plasma and liver (infer a short-term diet), while blood cells and muscle have a slower turnover rate and reflect long-term diet [24]. Some tissues such as feathers are inert after synthesis, and others such as hair or nails give a temporal sequence [22]. In addition, different elements have different rates of incorporation into the different tissue types. As an example of different incorporation rates, Ogden et al. [23] found, in captive Dunlin (a bird species), that in whole blood the half-lives for ^{13}C and ^{15}N were 11.2 ± 0.8 days and 10.0 ± 0.6, respectively.

Another study by Podlesak et al. [25] investigated the turnover of O and H isotopes in a group of woodrats. By feeding the animals isotopically consistent food, but by varying the isotopic concentrations in the water, they were able to determine the rate of turnover for O and H in the body water, hair and teeth for the woodrats [25]. They were able to determine the fractional proportion of atmospheric O_2, drinking water and food to the O and H makeup of the animals [25]. The body water turnover half-life was calculated to be 3 to 6 days, and it was estimated that O_2, drinking water and food were responsible for 30%, 56% and 15% of the O in body water, respectively [25]. Drinking water and food were found to be responsible for 71% and 29% of the H in body water, respectively [25]. The change in drinking water was also recorded in hair and tooth enamel [25]. Podlesak et al. [25] concluded that drinking water had a strong influence on the composition of body water, hair and tooth enamel, but that food and atmospheric O_2 also contribute O and H to the tissues. Other studies of turnover rates of different isotopes in different animals and different tissues have been published (for example [26–32]) but are beyond the scope of this chapter.

These differing rates of incorporation of different isotopes into different tissues make it obvious that the type of tissue will have an impact on the results and should be considered when the tissue type is known. It is also important to consider the fractional proportion of atmosphere, food and water to the isotopic makeup of the animal being studied. Further, it is important to consider the biology of the animal in question. For instance, animals from northern climates often have a winter and a summer coat, so any information obtained from the hairs of one of these animals will only be from, at most, one-half of the year. Mammals living in temperate climates, on the other hand, will not have any seasonal variation in their hair; however, wet and dry seasons can lead to a change in diet that can be detected.

We will now discuss the use of isotopes in forensic wildlife crimes and conservation. Several case examples will be presented and the relevant citations will be provided. In addition, cases involving human samples (box 6.2) and plant samples (box 6.4) are given as examples of the potential for the analysis of isotope ratios in other areas of forensic science.

Box 6.2 Human Case Examples

Stable isotope ratio analysis has been used to assist in the identification of humans as well as animal samples. When there are victims and no knowledge of whom they might be, determining where they come from can give investigators a starting point to determine their identity.

In 2002 a body of a male was found near a motorway in Germany [1]. The victim was beaten and shot and found in an advanced state of decomposition, such that no fingerprints could be obtained [1]. Other clues, such as dental work and clothing, pointed to the victim's being from Eastern Europe, possibly Romania [1]. Hair, teeth, bone and representative soil samples of the burial area were analysed for the isotope ratios of H, C, N, strontium (Sr) and lead (Pb) [1]. The Pb isotope ratio was found to be high, at levels found in places in North America, but expertise about the dental work precluded North America as a country of origin. After further investigation it was found that the Pb-isotope ratio of the teeth and skull contained Pb-sulphides, which are found in a large gold and silver mine in Romania [1]. This allowed police to look for missing persons in Romania, where they found a family who had filed a missing persons report. The victim's identity was confirmed with DNA [1]. After positive identification of the victim, the investigation proceeded and two people were arrested and subsequently confessed to the murder [1].

In 2005 the body of a male was discovered in Dublin, Ireland. The body was dismembered and decapitated; the head was never recovered [2]. Fingerprints were obtained but did not correspond with anyone in the database, and no other information was available to help determine the identity of the victim [2]. Stable isotope analysis was undertaken on four elements (^{2}H, ^{13}C, ^{15}N and ^{18}O) to aid the investigation and identification of the victim [2]. The victim's hair and nails were analysed, giving a history of approximately 200 days prior to death [2]. Based on these results and comparison to a control person, it was concluded that the victim had lived in or around Dublin for at least seven months prior to death [2]. The $\delta^{18}O$ signature from the outer femur (and hence the oldest bone) was indicative of only five regions worldwide, and by analysing the inner femur, it was determined that the victim moved to Dublin approximately 6.3 ± 2.7 years prior to his death [2]. This allowed a timeline and the region of origin was narrowed to the Horn of Africa, which allowed police, in conjunction with other evidence, to obtain permission to proceed with DNA tests (these would not have been approved without the conclusions of the isotopic analysis), which led to the positive identification of the victim [2]. Consequently, two females were arrested and convicted of the murder [2].

References

1. Rauch, E., S. Rummel, C. Lehn and A. Buttner, Origin assignment of unidentified corpses by use of stable isotope ratios of light (bio-) and heavy (geo-) elements—A case report. *Forensic Science International*, 2007, 168(2–3): 215–218.
2. Meier-Augenstein, W. and I. Fraser, Forensic isotope analysis leads to identification of a mutilated murder victim. *Science & Justice*, 2008, 48(3): 153–159.

One problem encountered in forensic science is the question of the exact type of species, as well as the region of origin of a sample. A good example of this problem is that of elephant ivory, especially if it has been carved into ornaments so gross morphological characteristics are removed. There are two genera of elephant: *Elephas*, the Asian elephant (*E. maximus*), and *Loxodonta*, the African elephant, of which there are two sub-species (*L. africana*, the bush elephant, and *L. africana cyclotis*, the forest elephant). All populations of *E. maximus* are listed in Appendix I of CITES, and their trade is subject to international laws. In contrast, *L. africana* is listed in Appendix I of CITES for most countries, but the populations of Botswana, Namibia, South Africa and Zimbabwe are listed in Appendix II. This means that trade in *L. africana* from certain populations fall under different laws from those of populations outside Botswana, Namibia, South Africa and Zimbabwe.

One means of differentiating between the two types of elephants has been developed by Singh et al. [33]. They use a multitude of tests to differentiate between African and Asian elephant ivory. ICP-MS and IRMS were used to determine trace metals and the difference in isotope ratios between the two samples. In this example, ICP-MS was used to analyse trace metals: cobalt (Co), molybdenum (Mo), rubidium (Rb), zirconium (Zr), chromium (Cr), copper (Cu), nickel (Ni), vanadium (V), lanthanum (La), cerium (Ce), praseodymium (Pr), neodymium (Nd), samarium (Sm), europium (Eu), niobium (Nb), thorium (Th), gadolinium (Gd), terbium (Tb), dysprosium (Dy), holmium (Ho), erbium (Er), thulium (Tm), ytterbium (Yb), lutetium (Lu), scandium (Sc) and yttrium (Y). Asian ivory was found to contain higher concentrations of V, Sm, Eu, Gd and Sc, whereas Mo and Ni were found at higher concentrations in the African ivory. Asian ivory was found to contain some rare earth elements, with the most discriminating being Y, which was found in all Asian samples but in none of the African samples.

The $\delta^{13}C$ and $\delta^{15}N$ values were also determined and compared. The $\delta^{13}C$ value was found to be unsuitable for separating the two geographic regions, as the standard deviations overlapped. However, the $\delta^{15}N$ value was found to be useful in separating the Asian and African elephants. The $\delta^{15}N$ is higher in African ivory than in Asian ivory, with no overlap of the standard deviations. Further, the $\delta^{15}N$ shows that different states can be separated into different clusters, for different regions, except for two of the regions, which had overlapping means. This would allow the determination of the species as well as the specific geographic origin of most ivory samples.

Other questions of origin include where an animal was caught or killed. This is especially relevant when dealing with animals that have large home ranges or migrate between the borders of different countries. One way to do this would be to look at the different isotope ratios found in the hairs of those animals and use them as a chronological clock of their movements. We will again look at the example of elephants and use hairs to trace their movements. Although it may be unlikely that elephant pelts are being sold or traded, the examples of the research by Cerling et al. in 2004 [34] and 2006 [35] give a good example of how this technique can be used for forensic wildlife purposes. It is also evident that, by the same methods and techniques, animals such as *Ovis canadiensis* (whose Mexican population is listed in Appendix II of CITES, but no other wild population is protected) can be assigned to a specific geographical region if encountered in an investigation.

In 2006, examining *L. africana*, Cerling et al. [35] used stable carbon isotopes ($\delta^{13}C$) in hair to determine dietary changes between browsing on trees and shrubs (C_3) and grasses (C_4) (see box 6.3 for the definition and differentiation of C_3 and C_4 photosynthesis). There is a large difference in the $\delta^{13}C$ ratio between plants using the C_3 and C_4 photosynthetic pathways [35]. In their study, Cerling et al. [35] examined the growth rates of $\delta^{13}C$ and $\delta^{15}N$ ratios in hair that they had collected between 2000 and 2002. The $\delta^{13}C$ and $\delta^{15}N$ ratios were also sampled for surrounding plant material. They focused on the resident population of 35 elephants from a national park and on one migrant elephant that visited the park several times per year [35]. Differences in the isotope ratios indicated rapid migration over long distances in the migrant elephant, but not in the resident elephant population [35]. They also determined there were differences in the C_4 fraction of the migrant elephant's diet, which could be attributed to crop raiding and could be quantified using stable isotopes [35]. Movements of some of the elephants were confirmed using GPS tracking [35].

Box 6.3 C_3 and C_4 Photosynthesis

Atmospheric CO_2 is assimilated by plants during photosynthesis [1]. C_3, C_4 and CAM are terms that refer to the photosynthetic pathways used by different types of plants. C_3 photosynthesis is known as the Calvin cycle [2], C_4 photosynthesis as the Hatch-Slack cycle [3], and CAM photosynthesis as Crassulacean acid metabolism [4]. Photosynthesis is the process by which, using energy from the sun, plants convert CO_2 and H_2O into energy. The process is carried out in the chloroplasts (the cellular organelles that give plants their green colour). The chloroplasts trap sunlight energy in chlorophyll molecules and use this energy to manufacture energy-rich sugar molecules [5]. During this process oxygen is released as a waste product [5]. The simplified photosynthetic equation is

$$6\ CO_2 + 12\ H_2O + \text{light energy} \rightarrow C_6H_{12}O_6 + 6\ O_2 + 6\ H_2O$$

However, this is a very simplified version of the whole process and there are many steps involved from start to finish. It is at one of these steps that C_3 photosynthesis and C_4 photosynthesis are differentiated. Whether plants assimilate CO_2 via the C_3 or C_4 photosynthetic pathway will determine the extent of preferential amplification of the lighter, ^{12}C, isotope [1].

C_3 photosynthesis is the simplest and least efficient form of carbon fixation used in plants; however, C_3 plants are better able to adapt to an increase in CO_2 by increasing their rate of photosynthesis [6]. They are associated with cooler and wetter climates and account for greater than 85% of the plant species on Earth and include all trees [1, 6]. C_3 plants account for most of the natural vegetation in temperate regions, high-altitude regions and tropical forests [1]. Barley, cotton, fallow grasses, legumes, rye and wheat are all C_3 plants. C_3 plants can lose up to 97% of their water, taken up by the roots from the soil, to transpiratory water loss [7]. They are called C_3 plants because their first products in the photosynthesis pathway after CO_2 are carboxylic acids formed of three linked C atoms [6, 8]. Primary CO_2 fixation involves a large isotopic effect and is catalysed by ribulosebiphosphate-carboxylase [9]. The first product of the Calvin cycle is 3-phosphoglycerate (a three C compound) [2, 9]. C_3 photosynthesis generally results in lighter isotopic $\delta^{13}C$ values in the range of -22‰ to -30‰ [1].

C_4 plants account for fewer than 5% of floral species [1] and have a high photosynthetic ability in warm climates with low CO_2 concentrations;

they dominate the tropical savannahs and grasslands [1, 8]. C_4 plants include sugarcane, maize and sorghum [1, 8]. Unlike C_3 plants, the first products of C_4 plants are four carbon acids, which boost photosynthesis in hot conditions [8]. Primary CO_2 fixation is catalysed by phosphoenolpyruvate-carboxylase, which results in a less pronounced isotopic effect, and the primary product is oxaloacetate, a four C compound [3, 9]. Heavier isotopic $\delta^{13}C$ values are found in C_4 plants in the range of -10‰ to −18‰ [1].

The third type of photosynthesis, adapted to arid conditions, called Crassulacean acid metabolism (CAM), which can operate under both the C_3 and C_4 photosynthetic pathways [1]. CAM was first discovered in a Crassulacean plant known as *Bryophyllum calycinum* [4] and is adopted in drought-tolerant species [8]. In CAM, the CO_2 is stored as an acid before use in photosynthesis. This type of photosynthesis is found mostly in succulents (cacti and other plants with water-storing tissue), some orchids and bromeliads (e.g., pineapples). In CAM photosynthesis the stomata* are only opened at night, when evaporation rates are at a minimum. The CO_2 is converted to an acid until daylight, when it is broken down and used in the photosynthetic pathway. C_3, C_4 and CAM photosynthetic pathways all differ in their isotopic fractionation, meaning that they all show different carbon isotope ratios.

References

1. Philp, R. P. and E. Jardé, Application of stable isotopes and radio isotopes in environmental forensics, in *Introduction to Environmental Forensics*, B. L. Murphy and R. D. Morrison, Editors. 2007, Elsevier Science & Technology: China, pp. 455–512.
2. Calvin, M. and A. A. Benson, The path of carbon in photosynthesis. *Science*, 1948, 107(2784): 476–480.
3. Hatch, M. D. and C. R. Slack, Photosynthetic CO_2-fixation pathways. *Annual Review of Plant Physiology*, 1970, 21(1): 141–162.
4. Ranson, S. L. and M. Thomas, Crassulacean acid metabolism. *Annual Review of Plant Physiology*, 1960, 11(1): 81–110.
5. Alberts, B., D. Bray, A. Johnson, J. Lewis, M. Raff, K. Roberts and P. Walter, Introduction to cells, in *Essential Cell Biology: An Introduction to the Molecular Biology of the Cell*. 1998, Garland Publishing, Inc.: London, pp. 1–36.
6. Smithsonian Environmental Research Centre. *C_3 and C_4 Plants* [cited 2008 26 April]; available from: http://www.serc.si.edu/labs/co2/c3_c4_plants.jsp.

---

* Stomata are openings in the leaves through which plants take in CO_2 and release O_2. Plants can control when stomata are open and how big the openings are. In hot and dry conditions it is preferable for the plants to keep their stomata closed to prevent water loss.

7. Raven, J. A. and D. Edwards, Roots: Evolutionary origins and biogeo-chemical significance. *Journal of Experimental Botany*, 2001, 52(suppl 1): 381–401.
8. Osborne, C. P. and D. J. Beerling, Review. Nature's green revolution: The remarkable evolutionary rise of C_4 plants. *Philosophical Transactions of the Royal Society B: Biological Sciences*, 2006, 361(1465): 173–194.
9. Asche, S., A. L. Michaud and J. T. Brenna, Sourcing organic compounds based on natural isotopic variations measured by high precision isotope ratio mass spectrometry. *Current Organic Chemistry*, 2003, 7(15): 1527–1543.

Cerling et al. [35] were able to show that, by analysing plant material in the area where the elephants were thought to have been feeding, they could correlate the stable isotopes within that plant material to sections from the elephant's hair, thereby confirming their movements. In addition to the confirmation of movement, this technique could be used to show that an animal *did not* feed in, or come from, a particular area, which could be very important to a case, especially one of smuggling animals or wild caught animals being sold as captive bred.

Another use of isotopes in forensic wildlife crimes is determining the difference between captive bred and wild caught animals from the same area. In the United Kingdom (UK) some species of birds have protected wild populations, but it is legal to trade in those that are captive bred with the proper registration and licensing in place. The same applies to certain species throughout Europe, although there are still thousands of wild song-birds illegally trapped every year [36]. According to the Royal Society for the Protection of Birds [37], there were 66 reported incidents of taking, possessing or selling of wild birds (non-birds of prey) during 2006 in the UK. Of those cases reported, fewer were investigated and only 12 resulted in successful prosecutions [36]. The major problem with this type of crime is proving if a bird is wild caught or captive bred. Currently, expert witnesses examine feather wear to determine age and origin, but this is a subjective analysis and often other expert witnesses are employed by the defence to refute the prosecution witnesses [36]. This makes the crime of wild caught versus captive bred animals very hard to prove and even harder to successfully prosecute.

A recent study comparing the stable hydrogen isotope ratios of wild caught and captive bred songbirds was performed by Kelly et al. [36]. They tested the δ^2H values for two morphologically distinct subspecies of wild goldfinch, *Carduelis carduelis britannica* from the UK and *C. c. major* from Siberia, and captive bred canaries (*Serinus canaria*) [36]. Kelly et al. [36] found that there was a significant difference between the wild UK goldfinch samples compared to those from Siberia and that they could be distinguished based on the δ^2H values alone. In their analysis of captive bred birds, they noticed a much smaller variation in the δ^2H values and found that they could

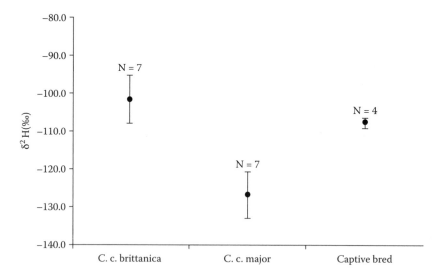

Figure 6.9 The δ^2H values measured in the primary feathers of two sub-species of goldfinch and from captive bred canaries (*Serinus canaria*). Error bars indicate 1 standard deviation. Obtained from [36] with permission.

not differentiate the captive bred birds from the wild bird population from the UK (figure 6.9).

Kelly et al. [36] concluded that it would be possible to identify illegally trapped songbirds using stable hydrogen isotope analysis if the areas where the birds were caught were widely separated geographically from where they were purportedly captive bred. They do feel that differentiation of captive bred and illegally trapped songbirds from the same geographical location will be more difficult [36]. However, their test could be expanded to include other elemental isotopes that could be used in combination with δ^2H to determine if a given bird was captive bred or wild caught. This approach has been used for other forensic samples for geographical determination in plants used for drugs (box 6.4).

Box 6.4 Isotope Analysis in Plants Used for Illicit Drug Manufacture

The analysis of isotopes has been used in other forensic investigations dealing with nonhuman samples, primarily to trace the synthetic pathway or identify the geographical origins of drugs [1–7]. The analysis of illegal drugs is often undertaken by forensic laboratories, and several common illicit drugs are produced from plants. Heroin is a derivative of morphine, which is synthesized from the opium poppy (*Papaver somniferum*). 'Magic mushrooms' refer to a group of fungi that contain psychedelic substances and which can be grown indoors (there are several species). Cocaine, the most widely used narcotic drug, is produced from the coca plant [3]. The ability to determine the region of origin of any of these drugs would aid police in tracing drug trafficking routes and potentially allow prosecution of individuals, or organisations, supplying multiple regions.

The potential to trace cocaine to a particular growing region has been investigated by Ehleringer et al. [3, 4]. Cocaine is an illicit drug derived from the plant *Erythroxylon coca*, commonly grown in South American locations. The ability to identify the region of origin of the original plant would be of major benefit for forensic investigations, as different seizures of the drug could then be traced back to, potentially, a common geographic origin and potentially a producer. In an initial study Ehleringer et al. [4] analysed 28 samples of cocaine from four growing regions (Bolivia, Colombia, Ecuador and Peru). They found that the four growing regions were isotopically distinct in their ^{13}C and ^{15}N isotope ratios, indicating the potential for the technique (figure B6.1) [4].

Expanding on their initial study, Ehleringer et al. [3] further determined the carbon (δ^{13}C) and nitrogen (δ^{15}N) isotope–ratio combinations for some of the coca-growing regions along the Andean Ridge. Using this ratio in addition to trace patterns in two alkaloids, they were able to

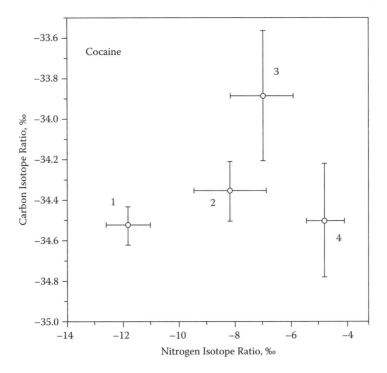

Figure B6.1 Carbon and nitrogen isotope ratios of cocaine samples originating from Bolivia (1), Peru (2), Ecuador (3), and Colombia (4). Error bars indicate 1 standard error from the mean. Obtained from [4] with permission.

correctly identify the regions of origin of 192 out of 200 cocaine samples (96 %) [3]. The isotope values for the coca leaves varied from −32.4‰ to −25.3‰ for $\delta^{13}C$ and from 0.1‰ to 13.0‰ for $\delta^{15}N$ [3]. Some of the regions tested could be discriminated solely by their $\delta^{13}C$ content [3]. Ehleringer et al. [3] proceeded to use bivariate (involving two variables, $\delta^{13}C$ and $\delta^{15}N$) mean and standard deviation parameters to determine the probability that a sample originated from a particular location.

Using the methodology described for the geographical profiling of cocaine, other samples, both plant and animal, could be assigned to a geographical location. This could aid investigations in many ways, including the identification of species protected in one region but not another and in determining wild caught and captive bred animals. Currently, other work is being carried out to determine if match sticks could be traced to a particular manufacturer based on isotope ratios (helpful for arson cases) [8, 9]. While this is slightly different from the theme of this text, the applications could easily be expanded to determine the regions of origin of CITES-listed wood or other materials.

References

1. Carter, J. F., E. L. Titterton, H. Grant and R. Sleeman, Isotopic changes during the synthesis of amphetamines. *Chemical Communications*, 2002, 2002(21): 2590–2591.
2. Palhol, F., C. Lamoureux, M. Chabrillat and N. Naulet, $^{15}N/^{14}N$ isotopic ratio and statistical analysis: an efficient way of linking seized Ecstasy tablets. *Analytica Chimica Acta*, 2004, 510(1): 1–8.
3. Ehleringer, J. R., J. F. Casale, M. J. Lott and V. L. Ford, Tracing the geographical origin of cocaine. *Nature*, 2000, 408(6810): 311–312.
4. Ehleringer, J. R., D. A. Cooper, M. J. Lott and C. S. Cook, Geo-location of heroin and cocaine by stable isotope ratios. *Forensic Science International*, 1999, 106(1): 27–35.
5. Billault, I., F. Courant, L. Pasquereau, S. Derrien, R. J. Robins and N. Naulet, Correlation between the synthetic origin of methamphetamine samples and their ^{15}N and ^{13}C stable isotope ratios. *Analytica Chimica Acta*, 2007, 593(1): 20–29.
6. Buchanan, H. A. S., N. NicDáeid, W. Meier-Augenstein, H. F. Kemp, W. J. Kerr and M. Middleditch, Emerging use of isotope ratio mass spectrometry as a tool for discrimination of 3,4-methylenedioxymethamphetamine by synthetic route. *Analytical Chemistry*, 2008, 80(9): 3350–3356.
7. Casale, J., E. Casale, M. Collins, D. Morello, S. Cathapermal and S. Panicker, Stable isotope analyses of heroin seized from the merchant vessel *Pong Su. Journal of Forensic Sciences*, 2006, 51(3): 603–606.
8. Farmer, N. L., A. Ruffell, W. Meier-Augenstein, J. Meneely and R. M. Kalin, Forensic analysis of wooden safety matches—A case study. *Science & Justice*, 2007, 47(2): 88–98.

9. Farmer, N. L., W. Meier-Augenstein and R. M. Kalin, Stable isotope analysis of safety matches using isotope ratio mass spectrometry—a forensic case study. *Rapid Communications in Mass Spectrometry*, 2005, 19(22): 3182–3186.

Stable isotope analysis has shown applicability to a wide range of disciplines within forensic science. The technique has been used to authenticate meats, milk and cheese (for example [38–41]) and for various migration studies (for example [42–46]), to name a few. However, the use of stable isotope analysis in forensic wildlife crimes has been limited and remains in its infancy. Based on the cases presented, it is obvious that the potential for this technology is starting to be realised. Isotope analysis is being used for more and more cases in the forensic science community.

This chapter has provided a very basic understanding of the theory and properties of the ICP-MS and IRMS. For a more detailed introduction to the technology of ICP-MS, a good place to start would be 'A Beginner's Guide to ICP-MS' by Robert Thomas (the complete series can be found in *Spectroscopy* 16[4] through 17[7]); for IRMS good descriptions can be found in references [1, 2, 20]. There are also textbooks entirely devoted to ICP (e.g., *Practical Guide to ICP-MS: A Tutorial for Beginners* by Robert Thomas, CRC Press; *Inductively Coupled Plasma Mass Spectrometry* edited by A. Montasser, Wiley-VCH) and to IRMS (e.g., *Modern Isotope Ratio Mass Spectrometry* by I. T. Platzner, Wiley, Blackwell). Further, a full list of isotopes can be found at http://physics.nist.gov/cgi-bin/Compositions/stand_alone.pl?ele=&all=all& ascii=ascii&isotype=all, from the National Institute of Standards and Technology (NIST).

Acknowledgements

Many thanks to Christine Davidson, Hilary Buchanan, Caroline Gauchotte, Vanitha Kunalan and Graham Reed for reviewing the manuscript and providing helpful advice.

References

1. Meier-Augenstein, W. and R. H. Liu, Forensic applications of isotope ratio mass spectrometry, in *Advances in Forensic Applications of Mass Spectrometry*, J. Yinon, Editor. 2004, CRC Press: Boca Raton, FL, pp. 149–180.
2. Benson, S., C. Lennard, P. Maynard and C. Roux, Forensic applications of isotope ratio mass spectrometry—A review. *Forensic Science International*, 2006, 157(1): 1–22.

3. Barrie, A. and S. J. Prosser, Automated analysis of light-element stable isotopes by isotope ratio mass spectrometry, in *Mass Spectrometry of Soils*, T.W. Boutton and S. Yamasaki, Editors. 1996, Marcel Dekker Inc: New York, pp. 1–46.
4. Coplen, T. B., J. K. Böhlke, P. D. Bièvre, T. Ding, N. E. Holden, J. A. Hopple, H. R. Krouse, A. Lamberty, H. S. Peiser, K. Révész, S. E. Rieder, K. J. R. Rosman, E. Roth, P. D. P. Taylor, R. D. Vocke Jr. and Y. K. Xiao, Isotope-abundance variations of selected elements (IUPAC Technical Report). *Pure and Applied Chemistry*, 2002, 74(10): 1987–2017.
5. Ebbing, D. D. and S. D. Gammon, *General Chemistry*. 6th ed. 1999, Houghton Mifflin Company: Boston, p. 1101.
6. Ehleringer, J. R., D. A. Cooper, M. J. Lott and C. S. Cook, Geo-location of heroin and cocaine by stable isotope ratios. *Forensic Science International*, 1999, 106(1): 27–35.
7. Sustainability of Semi-Arid Hydrology and Riparian Areas (SAHRA). Isotopes: Oxygen. 2005 [cited 1 May 2008]; available from: http://www.sahra.arizona.edu/programs/isotopes/oxygen.html.
8. Bowen, G. J. and J. Revenaugh, Interpolating the isotopic composition of modern meteoric precipitation. *Water Resources Research*, 2003, 39(10).
9. Ghazi, A. M., Application of laser ablation inductively coupled plasma mass spectrometry (LA-ICP-MS) in environmental forensic studies, in *Introduction to Environmental Forensics*, B. L. Murphy and R. D. Morrison, Editors. 2007, Elsevier Science & Technology: China, pp. 637–671.
10. Montaser, A., J. A. McLean, H. Liu and J.-M. Mermet, An introduction to ICP spectrometries for elemental analysis, in *Inductively Coupled Plasma Mass Spectrometry*, A. Montaser, Editor. 1998, John Wiley & Sons: New York, pp. 1–31.
11. Thomas, R., A beginner's guide to ICP-MS: Part II: The sample introduction system. *Spectroscopy*, 2001, 16(5): 56–60.
12. Kellner, R., J.-M. Mermet, M. Otto and H. M. Widmer, Elemental analysis, in *Analytical Chemistry*, R. Kellner, J.-M. Mermet, M. Otto and H. M. Widmer, Editors. 1998, Wiley-VCH: Berlin, pp. 433–526.
13. Kealey, D. and P. J. Haines, Inductively coupled plasma spectrometry, in *Analytical Chemistry*. 2002, BIOS: Oxford, pp. 209–213.
14. Thomas, R., A beginner's guide to ICP-MS: Part III: The plasma source. *Spectroscopy*, 2001, 16(6): 26–30.
15. Dean, J. R., A. M. Jones, D. Holmes, R. Reed, J. Weyers and A. Jones, Atomic spectroscopy, in *Practical Skills in Chemistry*. 2002, Prentice Hall: London, pp. 170–179.
16. Thomas, R., A beginner's guide to ICP-MS: Part IV: The interface region. *Spectroscopy*, 2001, 16(7): 26–34.
17. Thomas, R., A beginner's guide to ICP-MS: Part VI: The mass analyzer. *Spectroscopy*, 2001, 16(10): 44–48.
18. Skoog, D. A., F. J. Holler and T. A. Nieman, Mass spectrometers, in *Principles of Instrumental Analysis*. 1998, Saunders College Publishing: London, pp. 253–271.
19. Kealey, D. and P. J. Haines, Mass spectrometry, in *Analytical Chemistry*. 2002, BIOS: Oxford, pp. 209–213.

20. Asche, S., A. L. Michaud and J. T. Brenna, Sourcing organic compounds based on natural isotopic variations measured by high precision isotope ratio mass spectrometry. *Current Organic Chemistry*, 2003, 7(15): 1527–1543.
21. Meier-Augenstein, W., Stable isotope fingerprinting—Chemical element "DNA"?, in *Forensic Human Identification: An Introduction*, T. Thompson and S. Black, Editors. 2007, CRC Press: Boca Raton, FL, pp. 29–53.
22. Bowen, G. J., L. I. Wassenaar and K. A. Hobson, Global application of stable hydrogen and oxygen isotopes to wildlife forensics. *Oecologia*, 2005, 143(3): 337–348.
23. Ogden, L. J. E., K. A. Hobson and D. B. Lank, Blood isotopic (δ^{13}C and δ^{15}N) turnover and diet-tissue fractionation factors in captive Dunlin (*Calidris alpina pacifica*). *The Auk*, 2004, 121(1): 170–177.
24. Hobson, K. A. and R. G. Clark, Turnover of ^{13}C in cellular and plasma fractions of blood: Implications for nondestructive sampling in avian dietary studies. *The Auk*, 1993, 110(3): 638–641.
25. Podlesak, D. W., A.-M. Torregrossa, J. R. Ehleringer, M. D. Dearing, B. H. Passey and T. E. Cerling, Turnover of oxygen and hydrogen isotopes in the body water, CO2, hair, and enamel of a small mammal. *Geochimica et Cosmochimica Acta*, 2008, 72(1): 19–35.
26. Tieszen, L. L., T. W. Boutton, K. G. Tesdahl and N. A. Slade, Fractionation and turnover of stable carbon isotopes in animal tissues: Implications for δ13C analysis of diet. *Oecologia*, 1983, 57(1): 32–37.
27. Ayliffe, L. K., T. E. Cerling, T. Robinson, A. G. West, M. Sponheimer, B. H. Passey, J. Hammer, B. Roeder, M. D. Dearing and J. R. Ehleringer, Turnover of carbon isotopes in tail hair and breath CO_2 of horses fed an isotopically varied diet. *Oecologia*, 2004, 139(1): 11–22.
28. Bryant, D. J., P. L. Koch, P. N. Froelich, W. J. Showers and B. J. Genna, Oxygen isotope partitioning between phosphate and carbonate in mammalian apatite. *Geochimica et Cosmochimica Acta*, 1996, 60(24): 5145–5148.
29. Bryant, D. J. and P. N. Froelich, A model of oxygen isotope fractionation in body water of large mammals. *Geochimica et Cosmochimica Acta*, 1995, 59(21): 4523–4537.
30. Martínez del Rio, C. and R. Anderson-Sprecher, Beyond the reaction progress variable: The meaning and significance of isotopic incorporation data. *Oecologia*, 2008, 156(4): 765–778.
31. Gratton, C. and A. Forbes, Changes in δ^{13}C stable isotopes in multiple tissues of insect predators fed isotopically distinct prey. *Oecologia*, 2006, 147(4): 615–624.
32. Podlesak, D. W., S. R. McWilliams and K. A. Hatch, Stable isotopes in breath, blood, feces and feathers can indicate intra-individual changes in the diet of migratory songbirds. *Oecologia*, 2005, 142(4): 501–510.
33. Singh, R. R., S. P. Goyal, P. P. Khanna, P. K. Mukherjee and R. Sukumar, Using morphometric and analytical techniques to characterize elephant ivory. *Forensic Science International*, 2006, 162(1–3): 144–151.
34. Cerling, T. E., B. H. Passey, L. K. Ayliffe, C. S. Cook, J. R. Ehleringer, J. M. Harris, M. B. Dhidha and S. M. Kasiki, Orphans' tales: Seasonal dietary changes in elephants from Tsavo National Park, Kenya. *Palaeogeography, Palaeoclimatology, Palaeoecology*, 2004, 206(3–4): 367–376.

35. Cerling, T. E., G. Wittemyer, H. B. Rasmussen, F. Vollrath, C. E. Cerling, T. J. Robinson and I. Douglas-Hamilton, Stable isotopes in elephant hair document migration patterns and diet changes. *Proceedings of the National Academy of Sciences of the United States of America*, 2006, 103(2): 371–373.
36. Kelly, A., R. Thompson and J. Newton, Stable hydrogen isotope analysis as a method to identify illegally trapped songbirds. *Science & Justice*, 2008, 48(2): 67–70.
37. Royal Society for the Protection of Birds. BIRDCRIME 2006: Offences against wild bird legislation in 2006. 2006 [cited 2008 January]; available from: http://www.rspb.org.uk/Images/birdcrime%2006_tcm9-169702.pdf.
38. Renou, J.-P., G. Bielicki, C. Deponge, P. Gachon, D. Micol and P. Ritz, Characterization of animal products according to geographic origin and feeding diet using nuclear magnetic resonance and isotope ratio mass spectrometry. Part II: Beef meat. *Food Chemistry*, 2004, 86(2): 251–256.
39. Renou, J.-P., C. Deponge, P. Gachon, J.-C. Bonnefoy, J.-B. Coulon, J.-P. Garel, R. Vérité and P. Ritz, Characterization of animal products according to geographic origin and feeding diet using nuclear magnetic resonance and isotope ratio mass spectrometry: cow milk. *Food Chemistry*, 2004, 85(1): 63–66.
40. Boner, M. and H. Förstel, Stable isotope variation as a tool to trace the authenticity of beef. *Analytical and Bioanalytical Chemistry*, 2004, 378(2): 301–310.
41. Brescia, M. A., M. Monfreda, A. Buccolieri and C. Carrino, Characterisation of the geographical origin of buffalo milk and mozzarella cheese by means of analytical and spectroscopic determinations. *Food Chemistry*, 2005, 89(1): 139–147.
42. Hobson, K. A. and L. I. Wassenaar, Linking breeding and wintering grounds of neotropical migrant songbirds using stable hydrogen isotopic analysis of feathers. *Oecologia*, 1996, 109(1): 142–148.
43. Chamberlain, C. P., J. D. Blum, R. T. Holmes, X. Feng, T. W. Sherry and G. R. Graves, The use of isotope tracers for identifying populations of migratory birds. *Oecologia*, 1996, 109(1): 132–141.
44. Hobson, K. A., Tracing origins and migration of wildlife using stable isotopes: a review. *Oecologia*, 1999, 120(3): 314–326.
45. Atkinson, P. W., A. J. Baker, K. A. Bennett, N. A. Clark, J. A. Clark, K. B. Cole, A. Dey, A. G. Duiven, S. Gillings, P. M. González, B. A. Harrington, K. Kalasz, C. D. T. Minton, J. Newton, L. J. Niles, R. A. Robinson, I. D. L. Serrano and H. P. Sitters, Using stable isotope ratios to unravel shorebird migration and population mixing: A case study with Red Knot Calidris canutus, in *Waterbirds Around the World*, G. C. Boere, C. A. Galbraith and D. A. Stroud, Editors. 2006, The Stationery Office: Edinburgh, UK, pp. 535–540.
46. Wassenaar, L. I. and K. A. Hobson, Natal origins of migratory monarch butterflies at wintering colonies in Mexico: New isotopic evidence. *Proceedings of the National Academy of Sciences of the United States of America*, 1998, 95(26): 15436–15439.

Index